EARLY HISTORY
OF
HUMAN ANATOMY

EARLY HISTORY
OF
HUMAN ANATOMY

From Antiquity to the
Beginning of the Modern Era

By

T.V.N. PERSAUD, **M.D., Ph.D., D.Sc.,**
F.R.C.Path.(Lond.), F.F.Path.(R.C.P.I.), F.A.C.O.G.

Professor and Head, Department of Anatomy
Associate Professor of Obstetrics, Gynecology and Reproductive Sciences
University of Manitoba, Faculties of Medicine and Dentistry
Consultant in Pathology, Health Sciences Centre
Winnipeg, Canada

CHARLES   C   THOMAS   •   PUBLISHER
Springfield • Illinois • U.S.A.

Published and Distributed Throughout the World by

CHARLES C THOMAS • PUBLISHER
2600 South First Street
Springfield, Illinois 62717

© *1984 by* CHARLES C THOMAS • PUBLISHER
ISBN 0-398-05038-4
Library of Congress Catalog Card Number: 84-8479

Printed in the United States of America
Q-R-3

Library of Congress Cataloging in Publication Data

Persaud, T. V. N.
    Early history of human anatomy.

    Bibliography: p.
    Includes index.
        1. Anatomy, Human--History.   I. Title. [DNLM:
    1. Anatomy--biography.   2. Anatomy--history.
    QS 11.1 P466e]
    QM11.P47   1984   611'.009   84-8479
    ISBN 0-398-05038-4

For GISELA

PREFACE

ANATOMY is one of the oldest branches of medicine. Without a knowledge of it the diagnosis and treatment of diseases are inconceivable. This book portrays the early history of anatomy in relation to the practice of medicine. An attempt has been made to chart the momentous achievements and changing concepts from ancient speculations and philosophical notions to the dawn of the scientific era. I have tried to bring a human dimension to the subject by focusing not only on the celebrated anatomists of the past but also on some of the great scholars and their ideas against a background of both historical progress and human cultural development.

It would be impossible to write a book such as this without borrowing extensively from other sources. I have also visited numerous libraries, museums and scientific institutions in order to study rare works of interest in their collections. For all the help I have received, including permission to reproduce illustrations, I am greatly indebted.

At times, I felt I had taken on the impossible but my publisher, Mr. Payne E.L. Thomas, demonstrated infinite patience and gave the kind of encouragement that made the writing of this book a satisfying and exciting experience. Mr. Roy M. Simpson, Medical Photographer, prepared many of the photographs. My secretary, Mrs. Barbara Clune, not only meticulously typed the manuscript but also took a personal interest in helping me bring this work to completion.

T.V.N. Persaud

CONTENTS

Page

Preface vii

Chapter

1. ANCIENT RECORDS. 3
2. BEFORE HIPPOCRATES 29
3. HIPPOCRATIC CONCEPTS 33
4. ARISTOTLE 38
5. ALEXANDRIA. 44
6. DAWN OF THE ROMAN EMPIRE. 50
7. GALEN. 57
8. THE EARLY MIDDLE AGES. 70
9. MONDINO DE LUZZI. 89
10. LEONARDO DA VINCI. 101
11. MEDIAEVAL ANATOMISTS 114
12. ANDREAS VESALIUS — ARCHITECT OF THE NEW ANATOMY. 147
13. DE HUMANI CORPORIS FABRICA 161

Bibliography. 183

Index 195

EARLY HISTORY
OF
HUMAN ANATOMY

*"Die Anatomie gibt dem menschlichen Geiste
Gelegenheit das Tote mit dem Lebenden, das
Abgesonderte mit dem Zusammenhängenden, das
Zerstörte mit dem Werdenden zu vergleichen,
und eröffnet uns die Tiefen der Natur mehr
als jede andere Bemühung and Betrachtung"*

-Johann Wolfgang von Goethe

*"History is the ship carrying living
memories to the future"*

-Stephen Spender

CHAPTER ONE

ANCIENT RECORDS

PREHISTORIC PERIOD

FROM the spectacular fossil discoveries made by anthropologist Donald Johanson and his team in the Afar triangle of Ethiopia, man can now retrace his origins to less than 4 million years (Johanson and White, 1980; Lovejoy, 1981; Johanson and Edey, 1981, 1982). The most complete skeleton, named Lucy, was discovered in 1974 at a site called Hadar and is considered to be about 3.5 million years old. Fragments of thigh and skull bones found in the middle Awash River Valley, south of the Hadar, suggest a fossil hominid older than Lucy by more than 300 thousand years. These creatures walked upright, stood about 4.5 feet tall, and had brains somewhat smaller than those of chimpanzees. From the time of appearance of this creature, *Australopithecus afarensis*, many centuries must have passed before our prehistoric ancestors began to think and act in ways that would be considered to be intelligible. Exactly when this occurred remains a matter of speculation (Leakey, 1981; Lewin, 1983).

The Stone Age began about 2.5 million years ago and man's life during this early period in history has been dominated by hunting and the gathering of food. The abundant and extraordinary prehistoric paintings, engravings and reliefs found in different parts of the world emerged during the stone and ice ages and depict hunting scenes, animals and in some instances human figures (Obermaier and Kühn, 1930; Wendt, 1976; Hadingham, 1979; Leroi-Gourhan,

1982; Beltran, 1982, Sandelowsky, 1983). It is believed that this prehistoric cave and rock art evolved from hunting, myths and magic rituals. Undoubtedly the slaughtering of animals provided some crude anatomical insight, and wounds sustained by hunters might have given occasion for reflecting on the structure of the human body (Figs. 1-5).

Prehistoric paintings of 231 human hands were found in the cave

Figure 1. Venus of Willendorf. This palaeolithic limestone figurine (4 3/8″ tall) is believed to be one of the earliest known representations of the human form (25,000-20,000 B.C.). The head is almost faceless; the pendulous breasts and protuberant abdomen are symbolic of a fertility goddess (Natural History Museum, Vienna).

Figure 2. Female figurine of baked clay (4th millennium B.C., Arpachiyah provenance, by kind permission of the Trustees of the British Museum).

of Gargas, near Aventignan, France (Janssens, 1957); 114 of the hands revealed mutilation of one or more fingers, and in only ten cases were the hands complete. The pictures of the remaining hands have not been well preserved to determine whether they are intact or mutilated (Hooper, 1980). The age of the pictures has been estimated as possibly 30,000 years old, but the people it originated from, and the reason for the mutilation and for depicting the hands, remain a mystery. Hooper (1980) is of the opinion that these hand-drawings were deliberately executed and not the results of some accidental activity.

ANTIQUITY

The great civilizations of antiquity thrived because of the complex and efficient social organization and the technological, as well as cultural, advances that were achieved. Their real accomplishments have been largely neglected and still remain far from being fully appreciated because of the lack of accurate records and problems of deciphering the material that is now available.

An example of this in more recent times is the discovery of 16,500 tablets and fragments, written some 45 centuries ago, in the ancient city of Ebla which is located in present day Syria. The tablets and fragments were unearthed between 1974 and 1976 by a

Figure 3. Prehistoric rock painting of a woman; from the Brandberg Massif in Namibia. (Courtesy of Dr. Beatrice Sandelowsky; photograph by Robert Camby). This drawing, as well as the next two, were done in some shade of red or brown. Nothing is known of their masters, significance nor age. The rock art of Southern Africa is believed to have originated during the very beginning of the lower Stone Age and might have been executed by the ancient San population in order to depict their myths, customs and religious beliefs (see Sandelowsky, 1983).

team of Italian archeologists and the ancient inscriptions have sparked heated controversy regarding the origin of man and the source of his religions (Pettinato, 1981; Matthiae, 1981).

There are many surviving artifacts which clearly indicate that some anatomical representations have been made in prescientific times. In many respects these have paralleled the cultural evolution of man (Figs. 6-8).

Figure 4. Four individuals carrying equipment is depicted in this rock painting from the Brandberg Massif in Namibia (Courtesy of Dr. Beatrice Sandelowsky; photograph by Robert Camby). The artist has combined beauty with fantasy to achieve a unique simplicity in depicting the frolicking group.

Mesopotamia

Magic, sorcery and the practice of divination were distinct features of the great Babylonian civilization that flourished in the fertile valley between the rivers Tigris and Euphrates (Meissner, 1920-25; Dawson, 1930; Dhorme, 1949; Kramer, 1961, 1963; Mallowan, 1965). Here, where civilization probably began, clay tablets, several millennia old, were discovered which described monstrous births and the interal organs of sacrificial animals with the omens they predicted (Fig. 9).

The Mesopotamian diviners even made models of these organs for instructing their disciples. The liver and lung of the sheep were often used and different parts were carefully marked out with appropriate cuneiform scripts (Figs. 10 & 11) which were used for predicting the future and interpreting natural events (Jastrow Jr., 1914; Chiera, 1938; Grayson, 1980).

Egypt

The earliest of the Egyptian papyruses (Edwin Smith Papyrus) is probably a copy of one that was first written between 3000-2500

Figure 5. Prehistoric Namibian rock art depicting a male figure (Courtesy of Dr. Beatrice Sandelowsky; photograph by Dr. Robert Camby). Except for the anatomical misrepresentation of the right hand, other features appear realistic (see Wendt, 1976, and Sandelowsky, 1983, for origin).

B.C. It is essentially a surgical document and was translated and published just over 50 years ago (Breasted, 1930). Tumors, ulcers, abscesses, different wounds and fractures, as well as the prescribed treatment, which included suturing, cauterization, and the use of splints, are systematically listed in it. For the first time in recorded history, the word "brain" is mentioned, followed by a description of the gyri and meninges. The heart is mentioned as the center of a distributing system of vessels which pulsates (Major, 1954).

The papyrus purchased by Georg Ebers at Thebes in Egypt was written about 16 centuries before the Christian era, but was compiled much earlier (Fig. 12). The author is unknown, but the name of the famous priest and physician, Imhotep, has been suggested as a source (Leake, 1952). It was published as a facsimile edition with an introduction, commentary and notes in 1875. Although the major part of the Ebers papyrus deals with incantations, medications and

Figure 6. Three athletes exercising, (painted on a vase of Greek origin, 5th century, B.C.). Remarkable details of bodily movements and of the faces, hands and feet (by kind permission of the Trustees of the British Museum).

prescriptions for the treatment of diseases, the document contains numerous anatomical terms and made references to parts of the human body, so much so that it was considered to be the oldest known anatomical document (Macalister, 1898). It should, however, be pointed out that almost all of the eight discovered papyri, especially that of Ebers and Edwin Smith, mention the human body (Ranke, 1933; Grapow, 1935) and give suggestions for the treatment of various medical and surgical conditions.

The Edwin Smith papyrus, although dated about the same time as the Ebers papyrus, is also considered to have been copied from an ancient document as early as 3,000 B.C. However, the Ebers medical papyrus is complete and of the others considered to be the most

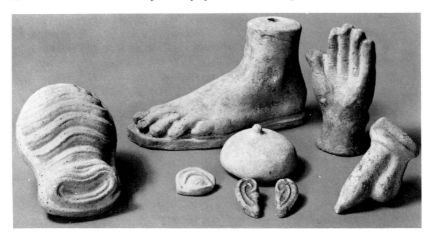

Figure 7. Terracotta models of parts of the human body (Roman, 3rd-1st century, B.C.). The uterus, left foot, hand, male external genitalia, ears, eye and breast are represented. Votive offerings (by kind permission of the Trustees of the British Museum).

important (Ebbell, 1937).

Interspersed with the description of various diseases are many anatomical terms (De Lint, 1932). Some of these annotations, however inaccurate, are of historical interest and several excerpts are presented from the translation that was made by Macalister (1898).

> There are vessels from it (the heart) to all the members. Each physician, master of healing, priest-exorcist, feels all these when he places his finger upon the head, upon the scalp (neck or occiput), upon the hands, upon the epigastrium, upon the arms or upon the legs. He traces all from the heart, because its vessels go into all his members, so he describes it (the heart) as the beginning of the vessels to all members.
> There are four vessels to the nostrils, of which two carry fluid (mucus?) and two carry blood. . . . There are four vessels to the sides of the temples, which if they carry blood to the eyes, all manner of diseases are produced in the eyes by their means, by their being open to the eyes. If water flows from them, it is the pupils of the eyes which give it. It is otherwise said that sleep causes it (the water?) to come from the eyes. . . .
> If the air enters into the nostrils, it is driven into the heart and (goes) through the intestine (by these vessels?), which distribute it to the whole body. . . .
> If excitement seized upon the heart, there is a rushing (of blood?)

Figure 8. Sculpture of *Ardha Nareeswar* (about 10th century A.D.) from a temple in Southern India. It depicts the half-male and half-female forms of *Ishwara*. The basic unit of man and woman is represented by the cosmic unity of *Shiva*, the creator, and *Shakti*, the progenitor of the vital force or energy. This dual and unique concept of the human form originated from the ancient Indo-Aryan civilization several millennia before the Christian era (courtesy of Professor R. Padmanabhan).

to parts of the intestine and to the liver. . . .
There are six vessels to the two arms — three to the right and three to the left, extending to the fingers. . . .
There are two vessels to the testes, which carry semen.

Figure 9. Clay tablet from the Royal Library of Nineveh (7th century B.C.). The cuneiform scripts are records of omens derived from examination of a sheep's liver used in divination (by kind permission of the Trustees of the British Museum).

There are two vessels to the kidneys, one to each kidney.
There are four vessels to the liver, to which they bring fluid and air. These give rise to all kinds of diseases when they (i.e. the fluid and air) are poured into the blood.
There are four vessels which extend to the anus, which supply it with fluid and air. . . .
There are over the anus, opening into it, two of these vessels — one on the right and one on the left — come to it from the leg. They cause dryness of the faeces.

During the period of the New Kingdom in ancient Egypt (late dynasty XVII through dynasty XX), a more accurate impression of the structure of the human body was probably first obtained because of the practice of embalming and mummification, which had reached its highest level at that time (Fig. 13). This ancient civilization believed in the immortality of the soul and resurrection of the

Figure 10. Babylonian clay model of sheep's liver (19-18 century B.C.) used in divination and for instruction in the temple. The cuneiform characters, inscribed in 55 sections, represent omens and magical formulae (by kind permission of the Trustees of the British Museum).

body.

After death only one of three spiritual elements departed, leaving two others, the *Ka* (his physical features and characteristics) and the *Ba* (soul). In death, the *Ba* travelled with the sun through the underworld during the night but returned to its resting place, the body, in the morning. It therefore became necessary to preserve the body with life-like appearances and to ensure that it is provided with everything conceivable that it might need for the next world. Thus, the funerary arrangements and furnishings evolved into a complex and elaborate ritual (Harris and Weeks, 1973; Andrews and Hamilton-Patison, 1978; Cockburn and Cockburn, 1980; Leca, 1981).

Most accounts of the practice of mummification are based on that recorded by the Greek historian, Herodotus, who visited Egypt in the 5th century B.C.

Figure 11. Inscribed clay model of the lung. It was used by barū priests for teaching divination to students in the temple, by comparison with the lung of a sacrificed animal (by kind permission of the Trustees of the British Museum).

Figure 12. Extracts (columns 1 & 2) from the Ebers papyrus (about 1600 B.C.). Although the document deals with incantations, medications and prescriptions, it contains many anatomical references (Karl-Marx-Universität Leipzig, Universitätsbibliothek).

There are a set of men who practice that art and make it their business. These persons, when a body is brought to them, show the bearers wooden models of corpses, painted so as to resemble

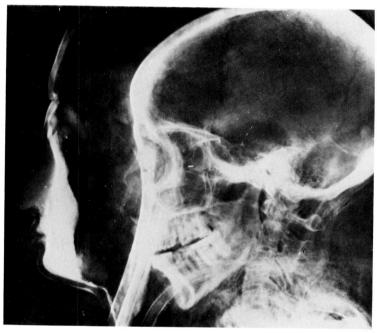

Figure 13. X-ray of the never-unwrapped mummy of Ta-pero, a court lady of the Twenty-second Dynasty, showing the perfect integration between the face and her sarcophagus (From James E. Harris and Kent R. Weeks, *X-Raying the Pharoahs*. Copyright © 1973, Charles Scribner's Sons. Reproduced with the permission of Charles Scribner's Sons.)

nature. The most perfect they say to be after the manner of Him whom I do not think it religious to name in connection with such a matter; they then show the second kind, which is inferior to the first and less costly; then the third which is the cheapest. All this the embalmers explain, and then ask in which way it is wished that the corpse should be prepared. The bearers, having agreed on the price, take their departure. The embalmers remain in their workshop, and this is their procedure for the most perfect embalming. First by means of an iron hook, they draw out the brain through the nostrils, taking it partly in this manner, partly by the infusion of drugs. Then with a sharp Ethiopian stone they make a cut along the whole contents of the abdomen, which they then cleanse, rinse with palm oil and rinse again with powdered aromatics. Then, having filled the belly with pure myrrh powdered, and cassia and every other kind of spicery except frankincense, they sew it up again. Having done this, they "cure" the body leaving it covered with natron for 70 days; it is forbidden to "cure" it for a longer space of time. At the expiration of the 70 days they wash the corpse and wrap the whole body in bandages of linen cloth, smeared over with gum, which the Egyptians commonly use in place of glue. After this the relatives, having taken the body back again, have a wooden case made in the shape of a man, and when it is ready, enclose the body in it. They fasten the case and store it thus in a sepulchral chamber upright against one of the walls. Such is the most costly way of preparing the corpse. When the middle style is chosen and great expense is to be avoided, they prepared the corpse in the following manner. They charge their syringes with oil from cedar and fill with it the abdomen of the corpse, without making any incision or taking out the bowels; they inject the oil at the fundaments and, having prevented the injection from escaping, they "cure" the body for the prescribed number of days, and on the last day they let out from the abdomen the oil of cedar they had previously injected; such is the power of the oil that it brings with it the bowels and the flesh, and nothing remains of the body but the skin and the bones. Having done this, they return the body without further operation.

The third method of embalming is the following, which is practiced in the case of the poorer classes: after clearing the abdomen with a purgative, they "cure" the body for the 70 days and deliver it to be carried (see Engelbach and Derry, 1942, cited in Harris and Weeks 1973).

It is now known that the process of mummification began before the early dynastic period, about 4 millennia B.C., among the inhabitants of the Nile Valley who believed in a life after death. The technique of mummification as described by Herodotus, and also 400 years later by another Greek historian, Diodorus Siculus, must have evolved to the level of perfection as practiced in the New Kingdom. More specific details of the embalming procedure, based on the painstaking research of Professor Albert Zaki Iskander, have been described by Harris and Weeks (1973).

Following death the body was taken to the house of mummification or to the house of purification. All clothing was removed and the corpse was placed on a large wooden board. The brain was removed through the nostrils, using a hooked wire. In rare cases, the brain was removed through a hole made in the skull.

The abdominal viscera, with the exception of the kidneys, were removed following an incision made in the left flank of the body. Following removal of the internal organs, the thoracic and abdominal cavities were washed with palm wine and spices. The internal organs were separately treated and placed in a container of natron for a period of 40 days.

Sprinkled with perfume and further treated with hot resin, the organs were wrapped in packages and placed in four canopic jars. The head of a deity was carved on the lid of each jar. After the Eighteenth Dynasty these were the Four Sons of Horus (Fig. 14).

In the Twenty-first Dynasty the embalmed and wrapped organs were returned to the body cavities, each accompanied by a wax figure of one of the Sons (Fig. 15). The wrapped organs were simply placed between the legs of the corpse in the Twenty-sixth Dynasty.

Figure 14. Canopic jars from Egypt. Twenty-first Dynasty (about 1000 B.C.). These small wooden containers represented the deities, the four sons of Horus, who were responsible for guarding the organs and viscera which were removed from the body during mummification and stored in them. From left: Duamutef (dog-headed) protected the stomach, Imsety (human-headed) protected the liver, Qebhsenuef (falcon-headed) was responsible for the intestines, and Mapy (ape-headed) protected the lungs. The heart as seat of the soul was left undisturbed in the body (by kind permission of the Trustees of the British Museum).

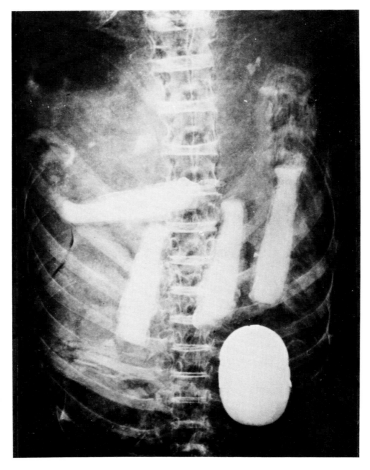

Figure 15. Frontal chest X-ray of Queen Nodjme (Twenty-first Dynasty) showing a large heart scarab and small wax statuettes representing the four sons of Horus. (From James E. Harris and Kent R. Weeks, *X-Raying the Pharoahs*. Copyright © 1973, Charles Scribner's Sons. Reproduced with the permission of Charles Scribner's Sons.)

The body cavities were filled with temporary stuffing in order to accelerate the dehydration process and to preserve the external features. The process of dehydration probably took half of the entire embalming period of 70 days as described by Herodotus. The packing material was removed from the body cavities, after which they were washed with water and palm oil, then dried. Resin or linen soaked with resin was then inserted in the cranial cavity and saw-dust, myrrh, and occasionally onions were packed into the abdominal cavity. After suturing the abdominal opening, the body was treated with a mixture of cedar oil, cumin, wax, natron, gum and possibly milk and wine, then sprinkled with spices.

In order to retain a life-like appearance, the cheeks were stuffed with linen and the nose was plugged. The sockets were also filled with linen and the eyelids were closed. A thick coat of molten resin was applied to the entire body in order to make the skin taut and to prevent loss of moisture. Eyebrows were painted on and often the body was decorated with jewelry.

The body was bandaged in a prescribed manner and everything that had been in contact with the mummy were then placed into six to seven large pots and buried near the tomb of the deceased.

Because of the crucial importance attached by the ancient Egyptians to the preservation of the human body after death in preparation for eternity, it comes as a startling surprise that there is no known depiction of the actual embalming process. Furthermore, there is no description of any of the internal organs nor any remarks of anatomical significance among the extensive archaeological remains from this exalted civilization.

In the spiritual and material context of life in ancient Egypt, preservation of the corpse was a practical ritual that simply ensured survival in the afterlife. It would seem that beyond this there was no other implication or importance as far as the embalmers or temple priests were concerned. Indeed, the Greek historian, Diodorus Siculus, remarked in his account of embalming, "then he who is called the cutter takes an Ethiopian stone, and cuts the flesh as the law prescribes, and forthwith escapes running, those who are present pursuing and throwing stones, cursing the defilement on to his head. For whosoever inflicts violence upon, or wounds, or in any way injures a body of his own kind, they hold worthy of hatred. The embalmers, on the other hand, the esteem worthy of every honour and respect, associating with priests and being admitted to temples without hindrance as holy men." (Cited in Harris and Weeks, 1973).

Mummification as practiced by the Egyptians was based on the religious conviction that life continues after death, an ancient belief that might have originated more than 50,000 years ago in the Neanderthal period. The well preserved head of a strangulated man, dis-

covered in the peat bogs of Tollund in Jutland and now in the Silke-
burg Museum in Denmark is considered to be that of a prehistoric
ritual murder.

China

Ancient Chinese medicine evolved along a unique course following the fundamental concepts of a complex balance between the two
forces, *Yin* and *Yang*, which determine the *Tao* (the way). Through
Buddhism medical knowledge flowed into China from India and later contacts were established with scholars from the Mediterranean
countries, particularly after the fall of the Roman Empire and the
expulsion of the Nestorian scholars from Constantinople.

Because of the doctrine of Confucianism dissection was not practiced in order not to defile the human body. Physicians were taught
the practical aspect of anatomy from models and diagrams until the
18th century. However, the medical scholars of ancient China revealed through their writings a keen sense of awareness of the human body for the treatment of diseases. The healing art in ancient
China probably began during the 4th millennia before the Christian
era by three legendary emperors (Hübotter, 1929; Huard and
Wong, 1968).

The I'Ching or "Canon of Changes," the greatest Chinese medical classic and probably the most ancient of Chinese books, is attributed to Emperor Fu Hsi. He probably was responsible for dividing
the universe and nature into the two governing cosmic principles,
Yang and *Yin*. Balance between these forces was considered to be the
basis for harmony and good health. Whereas the hollow organs and
viscera were *Yang* (light, moist and active), the solid ones were considered to be *Yin* (dark, dry and passive). Diseases resulting from external forces are *Yang* and those internally caused are *Yin*.

Emperor Shen Nung is regarded as the father of agriculture and
herbal medicine. Not only is he credited with the discovery of the
plough but also with the compilation of the first materia medica, the
Pen-Ts'ao or the great herbal. In three volumes Emperor Shen
Nung listed 365 drugs and their therapeutic benefits.

Huang Ti (2600 B.C.) is the father of Chinese medicine. He
compiled the Nei Ch'ing or the "Canon of Medicine" which dealt
with the functions and diseases of the human body. In his book,

Huang Ti made the remarkable assertion that "all the blood of the body is under the control of the heart. The heart is in accord with the pulse. The pulse regulates all the blood and the blood current flows in a continuous circle and never stops." He clearly recognized a relationship between blood, pulse and the heart which William Harvey (1578-1657) later confirmed.

Other anatomical-physiological annotations in the Nei Ch'ing were the following:

> The heart is a king, who rules over all organs of the body; the lungs are his executive, who carry out his orders; the liver is his commandant, who keeps up the discipline; the gall bladder, his attorney general, who coordinates and the spleen, his steward who supervises the five tastes. There are three burning spaces — the thorax, the abdomen and the pelvis — which are together responsible for the sewage system of the body.

The earliest medical writings are to be found in the Tso-Chuan, which was probably written about 540 B.C.

An interesting theory of intrauterine development was formulated by the physician Wen Tze about 300 B.C. He was particularly known for his interest in the diseases of women (Said, 1965).

> In the first month, the liquid becomes jelly-like. In the second month the veins and blood in them are formed. In the third month, the shape of human being is formed. In the fourth month, it takes the shape of an embryo. In the fifth month, the muscles become stiff. In the sixth month, the bones become hard. In the seventh month, the figure is complete and the soul enters. In the eighth month, some movement start with some agitation in the ninth month and early in the tenth month, the baby is born.

The great surgeon Hua T'o (?-208 A.D.) pioneered the use of anesthetics and is credited with the preparation of anatomical charts, the Nei-Chao-T'u, showing the organs of the body.

From the drawings of Yang Chiai (?1066-1140 A.D.) there was a renewal of interest in anatomy following a dissection in 1106 (Huard and Wong, 1968).

Taoism and the practice of acupuncture had an influence on anatomical thinking. The Taoist divided the human body into three regions: the upper part for the spirits, the connecting or middle part and, joined to this, the region of genital activity, represented by the paired kidneys. The drawings of the acupuncturist merely indicated the locations on the body where this procedure should be carried out.

During the Ming Dynasty, Wu T'uan (1438-1517 A.D.) edited his *Orthodox Medical Record* in 1515 A.D. This work was based on the information contained in over thirty other medical manuscripts, and listed fifty-one topics in need of clarification. Included were the principles of anatomy and physiology. It should be reiterated that even up to this period in the history of Chinese medicine the human body was considered to be so sacred that it was not dissected for anatomical studies. Even minor surgical procedures were not permitted, on account of reverence for the body, which in many respects retarded the progress of surgery.

The physicians of ancient China recognized that their body consisted of skin, flesh, muscles, tendons and bones. Nine orifices were identified: the eyes, ears, nose, mouth, anus, urethral and vaginal openings. With respect to the internal structures, they suggested that "man was composed of the five Tsang or storing organs and the six Fu or eliminating organs. The five Tsang were more important and they were the liver, heart, spleen, lungs and kidneys. The six Fu were the stomach, large intestine, small intestine, urinary bladder, gall bladder and the three burning spaces."

Of fascinating interest is a rare drawing of an ancient Chinese anatomy of unknown origin. It is in the form of a wood etching and measures approximately 74 × 24 cm showing the structure and functions of different parts of the body. It was probably done some time during the 16th century or even earlier. Said (1965) stated in his book that the English surgeon, Dr. Lockhart, remarked that "the Chinese pictures of anatomy look as if someone saw the incomplete dissection of the internal body and then has drawn the organs from memory, while he filled out the darker remaining parts from imagination and drew more what according to his own mind existed and not what actually existed." Undoubtedly, the lack of any practical knowledge of human anatomy, which during this period was only slowly evolving in Western Europe, hindered an accurate understanding of the structure of the body.

India

Like the Chinese, the Indian civilization is one of the oldest known. From the archeological excavations and findings of

Mohenjo-Daro and Harappa, it is now known that there were earlier inhabitants of the Indus Valley between the Himalaya and the Vindhya ranges, long before the Aryan conquest from the northwest in 1500 B.C. Abundant evidence suggests an advanced social organization and sanitation practice. The settlements were orderly laid out and there were wells, sewers and public baths.

With the cultural development inthis ancient Indo-Aryan civilization healing became entwined with religious practices (Sondern, 1936). Health and diseases were attributed to the Gods with Dhanvantari as the patron. Three "Doshas" or humours (wind, bile and phlegm) were thought to permeate the entire organism and when in harmony lead to good health. Notwithstanding their devotion to spiritual pursuits, the Aryans surprisingly developed a rational and secular approach to the practice of healing based on keen observations, the judicious use of herbs and surgery.

Several phases of development within this ancient culture have been identified: the Vedic (about 1500-500 B.C.), Brahmanic (600 B.C.-1000 A.D.) and the Mughal (from 1000 A.D. until 18th century). The Vedas or books of knowledge were compiled during the Vedic period. It has been suggested that these revealed works of the universal spirit or creator, which embodied religion and philosophy, were formulated some 4000 years prior to the Christian era. Only four of these books have survived: The *Rig Veda, Sama Veda, Yajur Veda* and the *Atharva Veda*. Many diseases and treatments, as well as surgical procedures, are recorded in two of these works.

The *Rig Veda* was essentially a medical treatise whereas the *Atharva Veda* (Fig. 16) was a surgical work. Thus, evolved traditional healing methods (Ayurvedic) together with practical skills which were remarkably advanced for that period in the history of man. Much of this knowledge spread into Asia and later reached Europe during the Middle Ages as a result of translations that were made by Persian and Arab scholars in the 11th century. For a deeper appreciation of the many remarkable accomplishments of ancient India in the field of medicine reference should be made to the work of Bhagvat Sinh Jee (1978) and to the authoritative and monumental twelve volume series edited by Singhal and Guru (1973-).

Apart from the use of medicinal plants for the treatment of a wide spectrum of diseases, the practice of surgery evolved to become one of the outstanding achievements of Indian medicine. The Laws

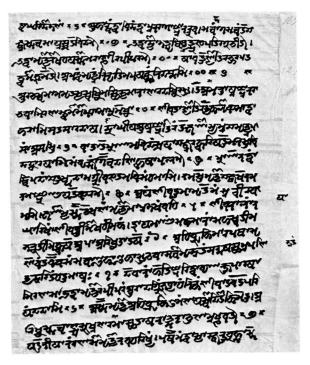

Figure 16. Page from the Paippalāda Atharva Veda (on birch-bark). One of the four "books of knowledge" from the ancient Indian civilization, it deals with philosophical, religious, magical and medical matters. (Universitätsbibliothek Tübingen.)

of Manu, which probably were formulated at about 3000 B.C. and compiled between 200 B.C. and A.D. 200, formed the basis of the social fabric of daily life in ancient India. The nose was cut off as a punishment for adultery and it is therefore not surprising that rhinoplastic procedures were well advanced. Other surgical operations included the repair of torn ear lobes and cleft lip, suturing of the intestine by applying large ants and followed by decapitation of the ants after they had bitten into the edges of the wound, removal of stones from the bladder and cataract extraction. As many as 101 surgical instruments, including a variety of forceps, scalpels, needles, and suturing material have been described. Aspiration of fluid for the treatment of both ascites and hydrocele and the use of a magnet for the extraction of foreign bodies were mentioned for the first time.

The famous physicians of Hindu medicine were Susruta, Charaka and Vagbhata. There is still somecontroversy as to exactly when these scholars lived. Susruta was a surgeon who most likely lived during the 6th century B.C. and taught at the University of Kasi or Banaras. He was a younger contemporary of Atreya who taught at Taksasila or Taxila, a famous seat of learning in the West.

The medical wisdom of Atreya was compiled in the form of a compendium or the Samhita. This work was essentially a classification of diseases with some remarks on the skeleton. Susruta also produced a similar Samhita but with more emphasis on surgical matters, including surgical instruments and surgical operations. It is in this work that one finds significant anatomical considerations of the ancient Hindu.

Because Susruta referred to Atreya's system of describing the bones, it is generally agreed that both these men lived and compiled their work during the 6th century B.C. In regard to Charaka it would appear that he flourished during the reign of King Kanishka, about the middle of the 2nd century. Vagbhata referred to both Susruta and Charaka by names and quoted their works. Vagbhata might have lived during the early part of the 7th century about 625 A.D. The work produced was a summary or Samgraha of the eight branches of medicine.

There is compelling evidence to believe that the knowledge of human anatomy revealed at the time of Susruta was gained not only by inspecting the surface of the human body but also through dissection of human subjects (Hoernle, 1907; Keswani, 1973). He recommended to those aspiring to a career in surgery that they should acquire a good knowledge of the structure of the human body and described the method how the body should be prepared for this purpose. Regarding the importance of dissection, he stated "therefore the surgeon, who wishes to possess the exact knowledge of the science of surgery, should thoroughly examine all parts of the dead body after its proper preparation" and about the method for dissecting the following is recommended (Singhal and Guru, 1973):

> Therefore for dissecting purposes, a cadaver should be selected which has all parts of the body present, of a person who had not died due to poisoning, but not suffered from a chronic disease (before death), had not attained a 100 years of age and from which the faecal contents of the intestines have been removed.

Such a cadaver whose all parts are wrapped by any one of "munja" (bush or grass), bark, "kusa" and flax, etc. and kept inside a cage, should be put in a slowly flowing river (Fig. 17) and allowed to decompose in an unlighted area. After proper decomposition for seven nights, the cadaver should be removed (from the cage) and then dissected slowly by rubbing with the brushes made out of any one of usira (fragrant root of a plant), hair, bamboo or "balvaja" (coarse grass). In this way, as previously described, skin, etc. and all the internal and external parts with their subdivision should be visually examined.

Susruta was able to achieve his remarkable knowledge of human anatomy in spite of religious laws which prohibited contact with the deceased other than for the purpose of cremation. Using a brush-type broom he was able to scrape off skin and flesh from the macerated remains in a systematic manner without actually touching the body.

Susruta's Samhita contains a fair amount of speculations and

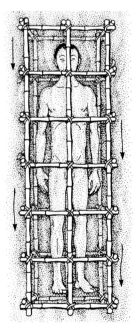

Figure 17. Preparation of a corpse for dissection as described by Susruta (see Singhal and Guru, 1973). Drawing by Glen Reid, Medical Illustrator, University of Manitoba.

philosophical concepts organized in a system of classifications. He stated that from surgical experience he knows of 300 bones, although 360 are recorded in the Vedas. He ascribed 120 bones to the extremities, 117 to the pelvis, flanks, back and the chest, and sixty-three to the region above the neck.

Susruta further described the types of bones, the importance of the bony skeleton, the number of joints, the types of joints, ligaments and muscles in different parts of the body. He assigned twenty additional muscles to the female on account of the breast and genital tract.

It has been suggested that Susruta might have arrived at the relatively large number of bones in the human skeleton because of the many dissections he carried out on children under two years of age. Despite his erroneous account of the skeleton and other speculations, e.g. 700 veins originating from the umbilicus and distributed to all parts of the body, Susruta's knowledge of human anatomy as revealed in his account of the muscles, joints, ligaments, and even blood vessels and nerves was remarkable for the period in which he flourished.

There has been speculation as to the extent of interactions between early Indian and Greek civilizations with respect to the evolution of medical knowledge (Sinh Jee, 1978; Hoernle, 1907, Major, 1954). Whether the Indian concept of the human body, in particular the bones, is based on a familiarity of Greek medicine, as is the case in the Talmud, is not easily resolved. Contacts had been established with Persian and Greek scholars as early as the 6th and 4th century B.C., respectively. According to Keswani (1973), Susruta's work was translated into Persian and Arabic in the 8th century and it was used as a surgical textbook (*Kitab-i-Susrud*) by students in the medical schools under the Caliphate.

There is no evidence that Susruta acquired his knowlege of medicine from the Greek school, but we do know that about 400 B.C. Skylax and Ktesias, both in the services of the Persian kings, visited India. Major (1954) believed that there must have been some exchange of knowledge between the two cultures contrary to the opinions of many scholars who have postulated an independent development with separate courses. He commented as follows:

> After the conquest of Alexander the Great, in the fourth century B.C., commerce with India was established, and Indian medical

science became part of the Greek heritage. Greek physicians became well acquainted with Hindu culture and medical science. Alexander, himself, had Indian physicians. In the later Greco-Roman period, it is obvious that Pliny and Galen borrowed heavily from Indian sources. During the Islamic period, Indian medicine was a powerful stimulus in the development of Arabic medical science. Harun al Raschid, the great caliph of Baghdad, in 773 A.D. called two Hindu physicians, Manaka and Saleah, to teach in the new medical school of Baghdad.

This controversial question of dependence and the sharing of medical knowledge among scholars of the ancient civilizations should be investigated further.

CHAPTER TWO

BEFORE HIPPOCRATES

ALCMAEON

ALCMAEON of Croton (Fig. 18), a contemporary of the mathematician Pythagoras and who lived in southern Italy about 500 B.C., is regarded as the earliest known person to have pursued actual anatomical studies. Very little of his writings have survived but it would appear that in addition to formulating theories in medicine and natural philosophy, he dissected animals with the sole purpose of understanding their anatomy (Codellas, 1932; Erhard, 1941; Sigerist, 1952; Arcieri, 1970; Lloyd, 1975).

Alcmaeon discovered the optic nerves and the pharyngotympanic tubes, which was again described by Eustachius (1520-74) in the 16th century. From a study of the developing chick embryo, he concluded that the head of the fetus was the first part to be formed.

Alcmaeon was one of the first to mention the brain and eyes. He asserted that the brain, not the heart, was the organ responsible for intelligence. Sleep he attributed to a transient suppression of cerebral blood flow which led to death when it became permanent.

The physiological studies carried out by Alcmaeon were aimed to unravel the nature of sense perception. He considered man to be the only creature that has understanding, whereas other animals "perceive but do not understand." He considered the senses to be connected with the functioning of the brain, so much so that if the brain moved or changed its position, the senses were incapacitated because the passages through which the sensations arrived were

Alcmœon

Figure 18. Alcmaeon of Croton (c. 500 B.C.).

blocked (Magner, 1979). Alcmaeon therefore inferred that the brain was the seat of intelligence.

In addition to describing the optic nerves, he postulated that external light, the "fire" in the eye, and the liquid in the eyeball were essential for vision. Undoubtedly, he was referring to the vitreous humor he had observed when dissecting the eyeballs of animals. It would appear that Alcmaeon recognized the importance of light entering the eye for vision to occur and the "fire" in the eye probably refers to the phenomenon following a severe blow (Codellas, 1932; Sigerist, 1952; Arcieri, 1970).

EMPEDOCLES

Empedocles of Agrigento in Sicily, a contemporary of Alcmaeon of Croton, lived between 493 and 433 B.C. He was a gifted philosopher and orator who speculated on the functions of the body. There are many legends attributed to Empedocles, including that of possessing supernatural gifts. He claimed to have had the power to heal the sick, relieve the aged of infirmities, to resurrect the dead, and to influence the rivers and the sun. However, when he was offered the governance of his city, he refused it. Fragments of Empedocles' two poems "on nature" and "purifications," highly philosophical work dealing with the nature of the universe and transmigration of souls, have survived (Van der Ben, 1975).

Empedocles carried out physiological studies pertaining to the sense organs, respiration, nutrition and growth (Siegel, 1959; Booth, 1960; O'Brien, 1970). He conceived four elements (earth, air, fire and water) as the "root of all things" (Longrigs, 1976) and also formulated the theory of the four humors, which formed the basis of humoral pathology. This profoundly influenced medical and scientific thinking for 2000 years until 1661 when Robert Boyle demonstrated otherwise in his *Sceptical Chymist*.

Empedocles believed that the life force was attributed to the "innate heat" of the body which was distributed by the heart. Such an hypothesis undoubtedly established the central role of the heart as part of the vascular system. It was the organ that distributed the "pneuma," the intangible life force that was more than both soul and life, throughout the body. The concept of the "pneuma" was expresssed a half century earlier by Anaximandros of Miletus (610-540 B.C.). For him air was the cosmic equivalent of the life-soul of man which he summarized as follows: "As our soul, being air, sustains us, so pneuma and air pervade the whole world."

Empedocles postulated several theories of embryological (Wilford, 1968) and physiological interest (O'Brien, 1970). Even though bizarre, he described the creation of many malformations, such as "faces without necks and arms without shoulders" as well as creatures "with faces and breasts on both sides, man-faced ox-progeny." He recognized the contribution of both mother and father to the formation of the embryo but suggested that its sex was determined by the degree of warmth within the uterus. He also thought that all living things inhale and exhale air through bloodless channels in the body and pores in the skin.

Such preoccupation with the "pneuma" and with the blood led another Greek philosopher, Diogenes of Apollonia (fl. 430 B.C.) to postulate his notion of the vascular system (Erhard, 1941; Cappelletti, 1975).

CHAPTER THREE

HIPPOCRATIC CONCEPTS

HIPPOCRATES

HIPPOCRATES (about 460-377 B.C.) was born on the Island of Cos in the Aegean (Fig. 19). He is considered to be the greatest of all physicians and the Father of Medicine, having established the healing art as a science far removed from superstition and magic. Relevant to our affluent and hurried times, Hippocrates advised that "diseases caused by overeating are cured by fasting, . . . diseases caused by indolence are cured by exertion . . . and tenseness by relaxation." Much of the so-called "Hippocratic Corpus," a large collection of philosophical, medical and scientific works, are wrongly attributed to him and were probably written by physicians of the medical school of Cos. Between 300-200 B.C., these writings were further compiled and edited by the scholars of the famous library at Alexandria (Adams, 1939; Jones, 1945; Wake, 1952; Stroppiana, 1963; Lloyd, 1975).

Of the collection of treatises only five are considered to be genuine. Except for the description of certain bones, what is mentioned of the structure of the human body is scanty, superficial, and largely inaccurate. Even though Hippocrates postulated that "anatomy is the foundation of medicine . . . and should be based on the form of the human body," he believed that one could learn sufficient anatomy from the observation of wounds without "the unpleasant if not cruel task" of dissecting corpses. Indeed, there is no evidence that Hippocrates dissected a human body and his descriptions were prob-

HIPPOCRATE,
Père de la Médecine

Figure 19. Hippocrates (about 460-377 B.C.). (Wellcome Institute Library, London.

ably made from inspection of the surface of the human body and chance observations of human wounds.

Hippocrates was apparently familiar with the bones of the skull and he knew of their articulations and sutures, but his knowledge of internal organs and of muscles was confused and speculative. He knew of blood vessels, but from the treatise, *On the Nature of Man*, one of the medical works of Hippocrates, a highly speculative description of the blood vessels in the human is presented, and sites for carrying out venesection recommended (Chadwick and Mann, 1950).

> The blood-vessels of largest calibre, of which there are four pairs in the body, are arranged in the following way: one pair runs from the back of the head, through the neck and, weaving its way externally along the spine, passes into the legs, traverses the calves and the outer aspect of the ankle, and reaches the feet. Venesection for pains in the back and loins should therefore be practised in the popliteal fossae or externally at the ankle.
>
> The second pair of blood-vessels run from the head near the ears through the neck, where they are known as the jugular veins. Thence they continue deeply close to the spine on either side. They pass close to the muscles of the loins, entering the testicles and the thighs. Thence they traverse the popliteal fossa on the

medial side and passing through the calves lie on the inner aspect of the ankles and the feet. Venesection for pain in the loin and in the testicles should therefore be done in the popliteal area or at the inner side of the ankle.

The third pair of blood-vessels run from the temples, through the neck and under the shoulder-blades. They then come together in the lungs; the right hand one crossing to the left, the left hand one crossing to the right. The right hand one proceeds from the lungs, passes under the breast and enters the spleen and the kidneys. The left hand one proceeds to the right on leaving the lungs, passes under the breast and enters the liver and the kidneys. Both vessels terminate in the anus.

The fourth pair run from the front of the head and the eyes, down the neck and under the clavicles. They then course on the upper surface of the arms as far as the elbows, through the forearms into the wrists and so into the fingers. They then return from the fingers running through the ball of the thumbs and the forearms to the elbows where they course along the inferior surface of the arms to the axillae. Thence they pass superficially down the sides, one reaching the spleen and its fellow the liver. Thence they course over the belly and terminate in the pudendal area.

Hippocrates used the term nerve to describe a tendon or sinew. The brain was considered to be a gland secreting mucous. The heart was described as a muscular organ of pyramidal shape and he placed within it the two auricles as reservoir of air and the two ventricles, the fountains of life, separated by a partition (Diller, 1938; Kapferer, 1951).

Hippocrates recognized the lungs, kidneys, the urinary bladder and the bowels. From the medical works attributed to him, we learn of a highly fanciful inter-relationship among these organs in certain disease conditions:

If a patient over the age of thirty-five expectorates much without showing fever, passes urine exhibiting a large quantity of sediment painlessly, or suffers continuously from bloody stools, his complaint will arise from the following single cause. He must, when a young man, have been hard-working, fond of physical exertion and work and then, on dropping the exercises, have run to soft flesh very different from that which he had before. There must be a sharp distinction between his previous and his present bodily physique so that the two do not agree. If a person so constituted contracts some disease, he escapes for the time being but, after the illness, the body wastes. Fluid matter then flows through the blood-vessels wherever the widest way offers. If it makes its way to the lower bowel it is passed in the stools in much the same form as it was in the body; as its course is downward it does not stay long in the intestines. If it flows into the chest, suppuration results because, owing to the upward tread of its path, it spends a long time in the chest and there rots and forms pus. Should the fluid matter, however, be expelled into the bladder, it becomes warm and white owing to the warmth of that region. it becomes separated in the urine; the lighter elements float and form a scum on the surface while the heavier constituents fall to the bottom forming pus.

Children suffer from stones owing to the warmth of the whole body and of the vesical region in particular. Adult men do not suffer from stones because the body is cool; it should be thoroughly appreciated that a person is warmest the day he is born and coldest the day he dies (see Chadwick and Mann, 1950).

POLYBUS

Of Hippocrates' many followers, his son-in-law, Polybus, of Cos (4th century B.C.), carried out studies on the human body and probably authored several of the "Hippocratic works" (Grensemann, 1968; Jouanna, 1969). He is described as a recluse who separated himself from the world and the pleasures it afforded in order to devote himself to the two treatises, *Nature of the Child* and *On Man*, which he compiled.

Polybus's anatomical representations were also crude and erroneous. In describing the major blood vessels of the body, he too mentioned four pairs which ran in extraordinary courses: from the head to the hips, lower extremities and ankle; the jugular vessels to the loins, thighs, leg and inner ankle; from the temples to the scapula and lungs, from there after mutual intercrossing to the spleen and left kidney, the liver and right kidney, and then to the rectum; and from the front of the neck to the upper extremities, the upper part of the trunk and the reproductive organs. This fanciful description of blood vessels was the result of pure philosophical speculation and emanated from the Hippocratic teachings.

HIPPOCRATIC LEGACY

In the Timaeus of Plato (429-347 B.C.), anatomical concepts as postulated by Hippocrates and Polybus have been reformulated to incorporate the doctrine of the macrocosm of the universe compared to the microcosm of the human body. His philosophical concept of the structure of the human body and causes of diseases profoundly shaped and directed the thinking of philosophers and natural scientists (Wright, 1925; King, 1954; Miller, 1962; Cornford, 1971).

From these Greek scholars emerged one whose systematic method of enquiry and careful observations of embryos and dissected animals laid the foundation of comparative anatomy. This was the great philosopher Aristotle whose many outstanding contributions in the biological sciences, politics and philosophy provide even up to this day a repository of knowledge which can only be admired.

CHAPTER FOUR

ARISTOTLE

PHILOSOPHER AND SCIENTIST

ARISTOTLE (Fig. 20) was the greatest natural philosopher of his era and, even though he was not a physician, his contributions to medicine have been equaled only by Hippocrates (Jaeger, 1948; Ross, 1952). Aristotle was born in 384 B.C., son of the court physician to King Philip of Macedonia, in the city of Stagira. He came under the influence of Plato and following the death of his teacher he travelled extensively and upon his return home he was requested by the King to tutor his son Alexander.

Returning to Athens in 355 B.C., Aristotle established his Lyceum which became a celebrated center for philosophical enquiries and the study of natural phenomena. He carried out extensive and fairly accurate studies, including dissections, on a wide variety of animals. In his work we see the fundamental concepts of organic evolution and Charles Darwin considered him to be the world's greatest natural scientist (Lloyd, 1968; Grene, 1972). Darwin remarked that Linnaeus and Cuvier were his gods, but compared to Aristotle they were mere school boys.

The Great Alexander, his former pupil, held him in the highest esteem. He often declared his gratitude to Aristotle for the cultivation of his mind and it was during the many expeditions of Alexander that valuable and rare specimens, both plants and animals, were collected.

Figure 20. Aristotle (384-322 B.C.) (Wellcome Institute Library, London).

ANATOMICAL WRITINGS

Like Hippocrates, Aristotle's knowledge of the human body was based on external observations and speculative ideas obtained from dissecting lower animals. Aristotle's diagram of the male urogenital system is probably one of the earliest known anatomical illustrations.

The four books on the *Parts of Animals (De Partibus Animalium)* and the first three books on the *History of Animals (Historia Animalium)* constitute a formidable volume of anatomical enquiries. Aristotle rightly described two main blood vessels located in front of the vertebral column and mentioned the aorta for the first time in history. He compared the thickness and consistency of the aorta and the vein and stated that they like all blood vessels arose from the heart and not from the head and brain as was previously stated by Polybus and others.

In describing the heart, Aristotle placed it in the center of the chest, located more on the left side above the lungs and near the bifurcation of the trachea ("so that it may counterbalance the chilliness of that side. For the left side is colder in man, as compared to the right"). The heart is considered to have a pointed apex which is more solid that the rest of the organ and directed forward. "In no animals does the heart contain a bone, certainly in none of those that we have ourselves inspected, . . . but it is abundantly supplied with sinews (tendinous fibres — chordae tendineae), as might be expected. For the motions of the body commence from the heart, and are brought about by traction and relaxation." Three cavities were described in the heart of "animals of great size": right, intermediate and left, with perforations leading into the lungs.

> Of these cavities it is the right which has the most abundant and the hottest blood, and this explains why the limbs on the right side of the body are also warmer than those of the left. The left cavity has the least blood of all, and the coldest; while in the middle cavity the blood, as regards quantity and heat, is intermediate to the other two, being however of purer quality than either (Ogle, 1882).

Aristotle described the great vessel (vein) as arising from the largest compartment on the right and the aorta from the intermediate.

The windpipe is described as located in front of the esophagus and mention is made of its construction out of a cartilaginous substance. The epiglottis is functionally described as "rising up during the ingress or egress of breath, and falling down during the ingestion of food, so as to prevent any particle from slipping into the trachea." In comparison to the animals Aristotle dissected, he described the human lungs as being anomalous because it was neither smooth nor divided into many lobes. He described the lung as large, rich in blood and spongy, like a foam. He thought that the "heat of the body" is cooled in the lung by external air during breathing, but not by the "innate spirit" as in "bloodless kinds." The bifurcation of the trachea is mentioned but it is considered to be united with the great vein and with the aorta. Aristotle believed that "those viscera which lie below the diaphragm exist one and all on account of the blood vessels; serving as a band, by which these vessels, while floating freely, are yet held in connection with the body. For the vessels give off branches which run to the body through the out-stretched structures, like so many anchor lines thrown out of a ship."

As far as the gastrointestinal tract is concerned, Aristotle gave a surprisingly accurate description of the esophagus as extending from the mouth, passing through the diaphragm and terminating in the quite distensible stomach. He described the human stomach as being like that of a dog and the lower part of the abdomen as being like that of a hog because it is wide. A fatty, broad and membranous mesentery extending over the bowels, as well as the large number of vessels in it, is mentioned. He identified the jejunum ("part of the small gut, of the gut, that is, which comes next to the stomach"), the cecum, the sigmoid flexure, and the rectum, but missed the duodenum. He knew of the spleen, the liver, the kidneys, and the bladder. His description of these organs is quite accurate.

The human spleen was described as being narrow and long, and the liver round like that of an ox. The presence of stones, growth and abscesses in the kidneys, the lung and liver are mentioned. He identified the porta hepatis and described veins in relation to it, but did not see any connection with the aorta.

Aristotle compared the kidneys to those of oxen and placed them close to the vertebral column, but with the right kidney at a higher level than the left. He identified two strong passages (the ureters) leading from the cavity of the kidneys to the bladder, and also two others leading to the aorta. Unlike the liver and spleen which "assist in the concoction of food," the kidneys take part in the separation of the excretion which flows into the bladder. In the center of the kidney, Aristotle identified "a cavity of variable size" and described the kidney as a solid organ, surrounded by more fat than other organs, and made up of numerous small kidneys. Except for the heart, Aristotle considered all organs to be devoid of blood. The blood vessels were supposed to contain air which is circulated and eventually cooled by the secretion of phlegm from the brain. The brain prevented overheating of the heart which was the seat of life and intelligence. Aristotle described the diaphragm with a central thin membranous (tendinous) and a peripheral fleshy (muscular) part. He viewed the diaphragm as "a kind of partition-wall or fence" which separated the nobler (thoracic cavity) from the less nobler (abdominal cavity) parts, where desire and grosser passions are located (Ogle, 1882).

Aristotle's description of the brain was quite interesting. He saw the brain in two portions enclosed by two membranes, the outer being the strongest. The human brain was considered to be the largest, in proportion to size, compared to other animals. Aristotle asserted that the brain "is larger in men than in women" . . . because the "region of the heart and of the lung is hotter and richer in blood in man than in any other animal; and in men than in women." He even reported finding more sutures in the skull of men than women because "the explanation is again to be found in the greater size of the brain, which demands free ventilation, proportionate to its bulk." He described a small cavity in the center of the brain linked by a membrane filled with veins but the brain itself was considered to be without blood because of which it was cold to the touch ("for of all the parts of the body there is none so cold as the brain"). Aristotle placed the cerebellum at the caudal end of the brain and described three passages leading from the eye to the brain, but based on simple inspection maintained that "it has no continuity with the organs of sense" (Clarke, 1963; Clarke and Stannard, 1963; Koelbing, 1968; Sorabji, 1970).

COMPARATIVE ANATOMY

Aristotle's vast contribution to anatomical knowledge was not based on human dissection but on speculation from the observations he made as a result of dissecting lower animals. In his work, *De Partibus Animalium*, he remarked that "the internal parts are not so well known, and those of the human body are the least known. So that in order to explain them we must compare them with the same parts of those animals which are most nearly allied."

Most remarkable were his embryological observations and his work on the *Generation of Man (De Generatione Animalium)* which he himself designated as his masterpiece (Platt, 1912; Preuss, 1970). In Aristotle's opinion, the embryo evolved from the egg even though the psyche, the principle of life, was derived from the male, and the womb merely serving as an incubator for the formless mixture of menstrual flow and semen which he described as "residual matter." His commentary as to how the child is formed, based on pure speculation and observation of the hen's egg, marks the beginning of embryological thinking (Morsink, 1982).

Aristotle, the most celebrated naturalist of antiquity, laid the foundation for comparative anatomy as a result of the systematic studies of animals and many dissections he carried out. Drawing from his observations he speculated about the structure of the human body (Lonie, 1964; Grene, 1972). Significant too was his influence on Alexander, the son of the King of Macedonia, later to be called Alexander the Great because of his great expeditions and conquests in the East. The most spectacular achievements in anatomy during this period (see Edelstein, 1935, 1967) were to be made in the city that was founded by Alexander on the banks of the river Nile.

CHAPTER FIVE

ALEXANDRIA

CULTURAL AND INTELLECTUAL CITY

ALEXANDRIA, the monumental city on the banks of the Nile was strategically located at the commercial crossroads of Asia, Europe and Africa. It evolved into a great cultural and intellectual center of the ancient world (Burn, 1982). Gathered here were eminent scholars who pursued with great vigor their interest in literature and in the sciences, encouraged by the ruling Ptolemies. With the fall of Greece to Rome the great minds of the period found refuge in Alexandria.

The library had an impressive collection of 700,000 volumes. It contained at that time all the learning of the civilized world. Together with the House of Muses or Museum, it attracted Jewish, Egyptian and Greek scholars who congregated here in the pursuit of knowledge (Parsons, 1952).

The embalmed body of Alexander the Great was kept in the mausoleum in the center of the city which was reported to be architecturally magnificent with its wide streets, temples, beautiful homes, and public buildings. The population was about 600,000 and consisted of Jews, Greeks, and Egyptians who lived in three different quarters of the city.

In this seaside cosmopolitan community the arts, philosophy and medicine flourished. In contrast to the dogmatism of the Aristotelian school, empiricism prevailed which was based on scientific investigations, actual observations, clinical histories and analogies.

ANATOMICAL STUDIES

The human body was dissected in order to understand more about its structure. The first anatomists credited with this distinction were Herophilus of Alexandria (Dobson, 1925; Potter, 1976) and Erasistratus of Ceos (Dobson, 1927; Lloyd, 1975), although this might not be entirely accurate in view of other evidence (see Singhal and Guru, 1973; Hoernle, 1907; Qatagya, 1982; Uddin, 1982). On the quiet banks of the Nile in a room that was used solely for anatomical studies, these two physicians made many anatomical discoveries through the dissection of a large number of cadavers that was made possible by their benefactors, Ptolemy Soter and Ptolemy Philadelphus.

It is said that Herophilus was the first anatomist to have gained a first hand knowledge of the actual structure of the human body and that he had dissected more human bodies than any of his predecessors. His extensive anatomical knowledge was attributed not only to the dissection of the dead, as many as 600 persons, but also of condemned criminals. According to Celsus (Scarborough, 1976), these were obtained "for dissection alive, and contemplated, even while they breathed, those parts which nature had before concealed."

Both Herophilus and Erasistratus have been accused of human vivisection. According to Tertullian as many as 600 living criminals were vivisected by Herophilus, and even fetuses were removed alive from the womb. Herophilus was described by Tertullian as "that doctor or butcher who cut up innumerable corpses in order to investigate nature and who hated mankind for the sake of knowledge." To this Celsus added "but to lay open the bodies of men whilst still alive is as cruel as it is needless" (see Ferngren 1982).

The reputation of these eminent physicians and teachers of anatomy attracted numerous pupils to the Alexandrian medical school. The achievement of Herophilus and Erasistratus are mentioned in the works of Celsus, Galen, Oribasius and others. Their works describing the human body on a factual basis have been lost.

HEROPHILUS

Herophilus was born about 300 B.C. and is often called the

"Father of Anatomy." Not much is known of his life and all of his writings, including his book *On Anatomy*, have been destroyed. He dissected extensively and made many anatomical discoveries as recorded in the works of other writers (Dobson, 1925; Potter, 1976). Because the Ptolemies encouraged dissection of the human body, Herophilus carried out as many as 600 dissections, both privately and for the public. Together with Erasistratus he was able to articulate two human skeletons which became widely known and attracted pupils from afar. It is also recorded that Herophilus was the first person to have tutored a woman as a physician. Whether he actually carried out human vivisection, as alleged by Celsus and St. Augustine, is not certain and appears unfounded.

Herophilus recognized the brain as the seat of intelligence, not the heart as postulated by Aristotle. He described the delicate arachnoid membrane of the cerebral ventricles which he considered to be the seat of the soul. Even to this day the confluence of the dural sinuses near the internal occipital protuberance is often called the torcular Herophili. Less known is the furrow in the floor of the fourth cerebral ventricle which he named the *Calamus Scriptorius*.

Herophilus recognized the lacteals, the functions of which he did not know. The nerves originating from the brain he considered to be the organs of sensation. He described the coverings of the eye and the name "duodenum" is attributed to him. The pulsation of arteries he attributed to the beating of the heart. He knew of the nature of the pulmonary artery and of the mesenteric vessels.

The discovery of the epididymis and of the lymphatic lacteals are attributed to Herophilus. From the work of Galen, his description of the human liver is recorded. He differentiated between nerves of sensation and those associated with voluntary movement and he knew that paralysis of muscles followed damage of these nerves. However, he used the word neuron to describe tendons and ligaments. The uterus was described in some detail by him and it has been anecdotally reported that among his pupils women were included, including Agnodice, who had to disguise herself as a man in order to practice medicine (Dobson, 1925; Souques, 1934; Potter, 1976).

ERASISTRATUS

Erasistratus (Fig. 21), a younger contemporary and rival of Herophilus, was born about 250 B.C. He argued that therapy should be directed to the local anatomical causes of diseases and rejected the concept that a knowledge of the entire body and its function in health was necessary for medical practice. In contrast, Herophilus, being a follower of the Hippocratic School of Medicine, emphasized treatment of the entire patient rather than individual symptoms.

Herophilus was a Dogmatist as opposed to the philosophy of Methodism practiced by Erasistratus. Even though he was an outstanding anatomist, Erasistratus considered himself a physiologist

Figure 21. The Greek physician Erasistratus is shown here reclining on a couch and conversing with an assistant. From an Arabic translation of Dioscorides' *De Materia Medica.* Iraqi Painting: A.D. 1224, Baghdad School, written by Abdallah ibn al Fadl. (Courtesy of the Freer ·Gallery of Art, Smithsonian Institution, Washington, D.C.)

and formulated experiments relating to bodily functions and diseases (Keele, 1961). His pneumatic theory (Wilson, 1959) was based on the flow of blood and two kinds of pneuma through minute channels of veins, arteries and nerves. The veins contained blood and the arteries distributed the vital spirit which was thought to be formed from air that passes from the lungs into the heart. When this vital spirit reached the ventricles of the brain it was transformed into the animal spirit which was then distributed throughout the body by the branches of nerves, conceived as being minute channels.

Erasistratus attributed all diseases to plethora, an accumulation of blood from food that remained undigested which obstructed the circulation of the vital spirit. His theory of the local accumulation of blood as a cause of diseases led him to direct greater attention to the heart, veins and arteries. He rightly saw the heart as a pump and suggested the existence of a very fine communication system between the arteries and veins. He described the auricles of the heart, a role for the semilunar and tricuspid valves. Regrettably, his pneumatic theory of the flow of vital spirit prevented him from understanding the true nature of the circulation of blood. He also described a large number of blood vessels, including the aorta, pulmonary artery and veins, hepatic arteries and veins, renal arteries, superior and inferior vena cava and the azygos vein.

Erasistratus recognized the function of the trachea. He differentiated the cerebrum from the cerebellum and noted the convolutions in different species of animals, with the greatest in man which he associated with the level of intelligence. The cerebral ventricles, which he considered with animal spirit, and the meninges, were also described by him. He saw the nerves as conveying animal spirit from the brain through tiny channels and perceived the contraction of muscles as the result of distention by animal spirit originating from the nerves (Dobson, 1927; Wilson, 1959; Lloyd, 1975).

DECLINE OF ALEXANDRIA

The decline of Alexandria is traditionally attributed to the invasion and conquest of the city, first by the Romans and later by the Arabs. It would appear, however, that the deterioration of Alexandria began even earlier because of internal politics, bickering and rivalry among the scholars.

Even during this distant period in history the value of basic research, including anatomy, was questioned as to its relevance in the treatment of patients. The followers of Herophilus clashed with those of Erasistratus. The Greek scholars were now being persecuted by the late Ptolemies, in particular the ninth Ptolemy (146-117 B.C.) who was known as the second Benefactor. The work of the medical scholars came to a standstill and began to deteriorate. By the 2nd century before the Christian era actual human dissections probably were not carried out, but some anatomical studies were pursued using animals.

CHAPTER SIX

DAWN OF THE ROMAN EMPIRE

DECLINE OF ANATOMY

FOLLOWING the persecution of the Alexandrian scholars and the conquest of Egypt by Caesar, Alexandria became part of the Roman Empire. The museum was destroyed and the library plundered and burnt. Persecution led to the dispersion of the Greek scholars within the great empire that was established by the Roman conquerors. With this began a new era in the evolution of medicine, and anatomy in particular, nurtured still by Greek scholars but culturally in the Roman environment. Egypt became incorporated into the Roman Empire during the reign of Augustus and even though the government was Roman, the culture was a hybrid and complex combination of both Greek and Roman. Religion, civilization and political power were fused.

The Romans were practical in their outlook and excelled in such areas as government, agriculture, medicine and warfare. The scientific pursuits and philosophical enquiry of the Greek scholars were held to be less important to the more immediate and practical goal of expanding the empire through conquest.

Medicine still remained the domain of the Greek physicians and about 60 B.C. a medical school, the "Asclepiadic sect in physic," was founded in Rome by Asclepiades of Bithynus (c. 120-30 B.C.). Anatomy suffered a decline because dissection was now forbidden and this situation did not change until the late middle ages (Kevorkian, 1959). Asclepiades did not regard anatomy as essential for the physician (Green, 1955) and even with the expansion of the medical school in Rome, first by Vespasian (A.D. 9-79) and others such as Hadrian (A.D. 117-138) and Severus (A.D. 208-235) dissection of the human body was not carried out although there is evidence that this might have been done up to the beginning of the second century of the Christian era (Fig. 22).

If this was not an era for scientific enquiry, then it was one for consolidating and documentation. The great recorders of the period were Lucretius, Caius Pliny Secundus, Cornelius Celsus and Marcus Tullius Cicero.

CICERO

Cicero (106-43 B.C.) considered superstition and dependence on religion as an obstacle for the establishment of proper institutions within society even though he was fully aware of the political importance of religion for citizens and their leaders (Fig. 23). In his book, *On Divination*, he eloquently objected to the dependence on religion for all matters from the pursuit of war to the administration of government. In his work, *De Natura Deorum*, he considered the nature

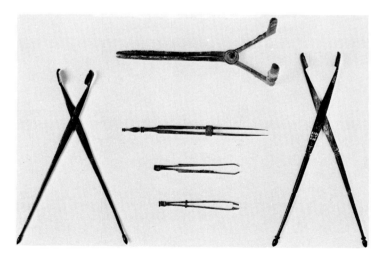

Figure 22. Surgical instruments from the early Roman era. (By kind permission of the Trustees of the British Museum.)

and origin of the universe and also presented a teleological account of the structure and function of the human body (Orth, 1925; Agrifoglio, 1961).

Lucretius (96-55? B.C.), like Cicero, objected to religion and superstition on the basis of the atomist philosophical concepts as taught by Democritus. His great poetic work *"De Rerum Natura"* is concerned with all living things and the universe for which he developed a natural explanation. His theory of origin of man, animals and plants, parallels in many respects Darwin's *Natural Selection*. Lucretius' philosophy, based on the teachings of Epicurus (341-270 B.C.) and of Democritus (460-370 B.C.), was purely speculative.

PLINY

Concerning Pliny (23-79 A.D.), the most famous Roman, naturalist and great encyclopedist, much has been written (Corney,

Figure 23. Terracotta (baked clay) votives (see also Fig. 7) of abdominal cavity with internal organs and tumor masses (Roman, 3rd-1st century B.C.). In ancient Greece and Rome it was customary to dedicate models of parts of the body at the shrines of the gods responsible for healing. (By kind permission of the Trustees of the British Museum.)

1914; Gudger, 1923). His expansive work, *Historia Naturalis*, consisting of thirty-seven books, attempted to deal with all of human knowledge in the arts and sciences (Fig. 24). Pliny stated in the preface that more than 20,000 facts have been selected from 200 books by more than 100 selected authors. Not only did he transcribe from innumerable works of Greek and Roman scholars, but he also recorded his own highly speculative philosophy and even the anecdotal statements made by others, however bizarre they might have been.

Despite the lack of any scientific merit in Pliny's astonishing interpretation of the natural world, his influence remained strong even up to the 17th century. According to Gudger (1923) "no other work contributed so much to keep natural history alive, . . . and following the appearance of the first printed edition in 1469, it was still the great authority, read, studied, and quoted by all students of natural history." Several books were devoted to the creation of the universe, the origin and description of man and animals, as well as strange mythological creatures such as the unicorn.

The fragments of anatomical statements in the work of Pliny, not unlike most of what he has recorded, were highly speculative and

Figure 24. Title page of Pliny's encyclopedical work. Basel edition 1554. (Courtesy of Dr. F.D. Bertalanffy.)

quite often erroneous. Most fascinating are his description of strange races of "wonder people" and "monstrous births." He mentioned Androgyni, combination of man and woman; Abermon, individuals with feet directed backwards; and Arimaspi with only one eye. The unicorn, an imaginary and fanciful creature, is described as a fierce animal with a body like that of a horse and long black horn emanating from its forehead. This ferocious animal was seen to have the feet of an elephant but the tail was that of a boar.

As a scholar of natural history, an overly enthusiastic recorder of historical and natural events, and a man of influence during the most glorious days of the Roman Empire, Pliny through his works discouraged the scientific study of the human body as practiced by the Greek scholars. A more influential contemporary of Pliny, whose expansive treatise in medicine contained many anatomical statements, was the Roman nobleman Cornelius Celsus.

CELSUS

Although not a physician, Celsus (c. 30 B.C.-45 A.D.) produced an encyclopedic work, *De Re Medicina*, which was a systematic survey of medical knowledge at that time fraught with particular emphasis on surgical practices (Fig. 25). He was probably a man of great learning because he wrote in impeccable Latin on a variety of subjects such as law, philosophy, and medicine. Most of his writings have been lost and his work on medicine was not discovered until the 14th century (Wellmann, 1913, 1924; Temkin, 1935).

De Medicina is in eight books, the last two of which deal largely with surgery and include a fair amount of anatomical descriptions, including a complete account of the skeleton. The work itself lacks originality and can be traced back to the writings of the Greek physicians. Nevertheless, it proved to be an important and influential book. Pope Nicholas V ordered its publication in 1478 and it was printed in Florence, being one of the first medical books to utilize the new process of moveable types (Orth, 1925).

The seventh and eighth books, as well as the introduction to the fourth book, contain many anatomical references and descriptions in relation to the treatment of diseases, in particular surgical conditions. Despite the lack of originality in his work, Celsus is rightly

A . CORN. CELSVS.
EX ICONIBUS A SAMBUCO EDITIS

Figure 25. Aulus Cornelius Celsus (Wellcome Institute Library, London).

considered one of the most outstanding medical writers of antiquity to have survived the ages (Castiglioni, 1940; Lipsett, 1961).

Celsus differentiated the windpipe from the esophagus and knew that the esophagus ended in the stomach. He gave an excellent description of the diaphragm, the liver, the spleen and the kidneys, but seemed to have been unaware of the duodenum because he thought that the stomach was directly connected to the jejunum by the pylorus. The duodenum was described earlier by the Alexandrian anatomist, Herophilus. Celsus knew enough about the urinary bladder, the uterus, and vagina to describe surgical procedures relating to these organs. Of greater merit, however, was his description of the skeleton. He described the features of the bones of the skull, including sutures and many of the openings. His description of the vertebrae, the ribs, the scapula, the humerus, radius, ulna, the tibia, fibula, the metacarpals and metatarsal bones is commendable. It would appear too that he was familiar with the semicircular canals of the inner ear and with the perforated features of the ethmoid bone (Orth, 1925; Agrifoglio, 1961).

Celsus recognized the importance of anatomy and even advocated human dissection. His description of surgical operations for the treatment of wounds, goiter, hernias, and cataracts revealed an appreciation for anatomy in carrying out these procedures. If human dissection was carried out during this period, there is no record of this in the work of Celsus and his knowledge of anatomy must have been obtained from the Greek medical scholars. He strongly opposed the alleged vivisection carried out by Herophilus and Erasistratus on condemned criminals (Scarborough, 1976). Indeed, he wrote "I regard it useless and cruel to open the living body, but it is necessary for those who study to see corpses, in order to learn the position and arrangement of the single parts, a thing that is seen much better in the cadaver than in the living body" (Castiglioni, 1941).

CHAPTER SEVEN

GALEN

MEDICAL SCHOLAR
AND CELEBRATED ANATOMIST

THE most celebrated anatomist of antiquity was the great physician Claudius Galen (Fig. 26). He was born in Pergamon, along the coast of Asia Minor, the son of a prominent Greek mathematician, architect and astronomer. His birth in A.D. 131 coincided with the period of greatest glory of the Roman Empire.

Galen studied philosophy under the influence of the Stoics, the Academicians, the Peripatetics and the Epicureans, and later medicine in accordance with a dream his father had received that his son should become a doctor. From the age of twenty, Galen travelled extensively in order to study with the greatest teachers of philosophy and medicine. He visited Greece, Palestine, Phoenicia, Crete and finally settled in Alexandria where he remained until he was thirty years of age. He was exposed to anatomy by Satyrus, a student of Quintus, and later to Pelops of Smyrna and later Heraclianus in the Alexandrian school (Green, 1951).

Galen's intellectual pursuits were such that his works were considered to encompass all of man's knowledge. His interests and writings went beyond medical matters and included philosophy, religion, mathematics and grammar. He was an extremely prolific writer (Ilberg, 1889, 1892, 1896, 1897; Walsh, 1934a, b; Peterson, 1977) who even produced a guide, entitled *On His Own Books*, to his voluminous treatises. In the field of medicine alone, Galen is credited with more

Figure 26. Claudius Galen (Wellcome Institute Library, London).

than 130 treatises which became the unquestionable repository of medical knowledge for more than one thousand years after his death in A.D. 201. Only eighty of these have survived. Nonetheless, Kuhn's edition of Galen's work took twelve years to complete and spans twenty formidable volumes (Kuhn, 1821-1833).

When he was twenty-eight years old, Galen returned to the city of Pergamon where he served as surgeon to the gladiators for four years. Probably, during this period he was able to make some anatomical observations as a result of attending to the wounds sustained by the gladiators. Human dissection was forbidden in Imperial Rome.

Galen derived anatomical knowledge from the works of his teacher and their predecessors, as well as from the extensive dissections he carried out on many species of animals. Galen dissected apes, monkeys, dogs, pigs, and even bears. He even considered the internal organs of man not to be very different from those of the pig. The voluminous writings of Galen contained many detailed anatomical descriptions (Withington, 1922) and there has been much debate over the years as to what constituted his personal observations and those of his predecessors.

Galen recognized the importance of anatomy for the work of the physician. He encouraged his pupils to visit Alexandria, so they might examine the human skeleton that was there, which he himself had studied. This is evident from his own work which contains the most detailed and systematic description of human bones (Singer, 1952). It is from the writings of Galen too that we obtain an insight as to the accomplishments of the early anatomists, particularly those who had lived in Alexandria.

Despite many errors and shortcomings, Galen's anatomical writings remained unchallenged up to the time of Vesalius in the 16th century. Rather than accepting the errors in the work of Galen, the French anatomist, Jacobus Sylvius (1478-1555) remarked in response to Vesalius that "then man must have changed his structure in the course of time, for the teaching of Galen cannot err." The Royal College of Physicians of London in 1559 even made one of its members, Dr. John Geynes, retract his statement that there were 22 inaccurate passages in the works of Galen. According to Clark (1964) "in the eyes of the College, his offence was double: indoors it was heterodoxy but out of doors it brought the chosen intellectual foundation of the medical art into question if not into contempt." Dr. Geynes was subsequently examined and admitted to the College in November 1560.

The influence of Galen continued for many decades and appeared in examination notices of the College. In 1595, Dr. Edward Jordan, a medical graduate of the University of Padua, was required to read five of Galen's works before being admitted to Fellowship. A Dr. Hood was forbidden to practise because he did not read Galen's work. He confessed that the books were too expensive to purchase. In 1596, Dr. Thomas Rawlins was failed "and was admonished to work harder, particularly at Galen" (Clark, 1964).

Galen derived his knowledge of anatomy largely from the dissection of animals, but in his own work, *On Bones*, he mentioned that he was able to study human bones from tombs that were destroyed. He also observed the bones of a body that had been washed from its grave by a flood and deposited on the river bank. He was able to observe the intact and articulated bones amidst the petrified flesh and the skeleton "lay ready for inspection, just as though prepared by a doctor for his pupils' lesson." And in another instance he studied the skeleton of a robber whose flesh was eaten away by birds and after two days it was ready "for anyone who cared to enjoy an anatomical demonstration" (Duckworth, 1962).

That there was no opportunity for dissecting the human body during Galen's time was probably due to the assumption that the structure of man and animals, in particular apes and monkeys, was fundamentally the same (May, 1968). The lack of enthusiasm for human dissection can be ascribed further to the prevailing hostility against such a pagan practice.

Galen himself marvelled at the complexity and delicate arrangement of the parts of the human body. According to the Aristotelian philosophy, he saw nature as providing a form and structure in the parts of the human body, with nothing being superfluous, in carrying out their functions. Such perfection and exquisite design can only be attributed to divine providence. It was therefore not surprising that Galen's account of the structure of the human body was accepted by Christians, Jews, and Moslems without questioning, and accounts for the fact that his works were translated into many languages and remained popular for more than 1,400 years. Even though Galen's real contribution to our knowledge of human anatomy were few, he still stands out as one of the greatest medical scholars not only of antiquity but of all time because of the profound influence his voluminous writings had on medical scholars through the Middle Ages (Temkin, 1973). Indeed, during this period the whole of medicine, and anatomy in particular, became stagnated and even suffered a decline.

In A.D. 161, Galen made his first visit to Rome. It was the beginning of the reign of the Roman Emperor Marcus Aurelius (A.D. 121-180), and following the suggestion of his patron, the Consul Flavius Boethus, he began to compose his anatomical works *On Anatomical Procedure* (Duckworth, 1962) and *On the Usefulness of the Parts of the Body* (May, 1968). He returned to Pergamon in A.D. 163 but later joined the military expedition of Marcus Aurelius in Aquileia as a personal physician. Because of the outbreak of the plague, the army returned to Rome and Galen was entrusted to care for the son of Marcus Aurelius, Commodus. For over twenty years (A.D. 169-192) Galen remained in Rome in pursuit of his medical work and in compiling his treatises. He returned finally to his home city in A.D.

192 where he died eight years later at the age of seventy.

ANATOMICAL TREATISES

Galen's life and anatomical accomplishments, as well as his other works, have been critically discussed by Prendergast (1930), Singer (1956), Sarton (1954) and May (1968). Sarton in his monograph provides an excellent introduction to the works of Galen, including those treatises that were translated from the Arabic and the more important editions that are available in English, including his work, *De Anatomicis Administrationibus*, which describes his anatomical studies in sixteen books. The first nine volumes were translated by Singer in 1956 from the surviving Greek text into English and the latter part of the work, the remaining seven books which had existed in Arabic, by Max Simon (1906) and Duckworth (1962). A complete translation from Greek into English of Galen's most influential work, *De usu partium (On the Usefulness of the Parts of the Body)*, was carried out by Margaret May and published in two volumes. It consists of seventeen books, devoted to both anatomy and physiology, which at the same time extolled the complexity and beauty of the human creation (May, 1968).

Sarton (1954) stated that Galen's treatises, *De Anatomicis Administrationibus*, or the *Anatomical Procedures* as it is now called was not known in the Middle Ages, but the Latin translation from the Greek text of the first nine books was carried out by Johann Guenther of Andernach and was first published in 1531 in Paris, several years before the publication of Vesalius' work. The remaining seven books became available only following its translation from the Arabic edition by Max Simon in 1906. Owsei Temkin translated portions of the German text into English and a more complete translation was achieved by Duckworth.

Some of Galen's other anatomical works included the following: one on the dissection of veins and arteries *(De Venarum Arteriarumque Dissectione)*; the dissection of nerves *(De Nervorum Dissectione)*; a *Secundum Naturam in Arteriis Sanguis Continator* (whether the arteries are naturally filled with blood); myology *(De Musculorum Dissectione ad Tirones)*. Some of his physiological treatises included a fair amount of anatomical information, often evolving from his dissections. *De Mus-*

culorum Motu deals with both voluntary and involuntary movements. His treatise, *De Usu Respirationis*, describes experiments on the pleura and the use of a bellows for maintaining artificial respiration.

ANATOMICAL STUDIES

Galen described the bones and sutures of the skull (Singer, 1952; May, 1968). He distinguished the squamous, the styloid, the mastoid, ethmoid, sphenoid, the maxillary, and the petrous part of the temporal bones. The malar bone was described as the zygomatic and he recognized the quadrilateral shape of the parietal bones. He described the essential features of the vertebral column, including the coccyx and the sacrum which he considered to be the most important. Twenty-four vertebrae were recognized and these were divided into cervical, dorsal and lumbar. His description of the ribs, the sternum, clavicle and of the bones of the extremities, as well as their articulations, was fairly accurate.

Even though Galen's description of muscles was based on the dissections he carried out on animals, it was fairly accurate and more advanced than that of any previous anatomist (May, 1968). Starting with the definition of what a muscle is (a bundle of fibers terminating in an independent tendon) he described and named individual muscles. These included the mylohyoid, the thyrohyoid, the six muscles of the eye, two muscles of the eyelids, four muscles of the lips, a muscle to each *ala nasi*, the frontal muscle and four pairs of muscles that move the lower jaw (the temporal, the masseter, the digastric and internal pterygoid muscles). In a similar manner, he systematically described the muscles of the tongue, neck, upper and lower limbs and the trunk. Despite the lack of adherence to any nomenclature, Galen's identification and description of the normal function of approximately three hundred muscles have been so successful that it survived to a large extent up to the present time.

In describing the brain, which he considered to be the seat of the soul, Galen identified seven pairs of cerebral nerves. Because they originated from the brain, they were considered to be nerves of sensation in contrast to the thirty pairs of spinal nerves which he recognized and designated as nerves of motion.

Galen did not recognize the olfactory and trochlear nerves. The optic nerve was designated as the first and both oculomotor and abducent were considered to be the second. He saw the trigeminal nerve as the third and fourth pair of nerves and his fifth pair of nerves was the facial and auditory combined. Galen named the glossopharyngeal, vagus, and spinal accessory nerves as the sixth pair of nerves and the hypoglossal was considered to be the seventh. He knew of the sympathetic nervous system and of the recurrent laryngeal branches of the vagus nerve on both sides of the neck. He described experiments involving the nerves that control movement of the tongue and voice production.

Galen has left us with a detailed description of the dissection of the brain of animals and the names of many structures he had identified are still with us, including the corpus callosum, the corpora quadrigemina, the fornix, the pineal body and the septum pellucidum. He knew of the ventricles, their communications, and of the choroid plexus. His description of the third and fourth ventricles reflected keen observations.

In describing the nerves, arteries and veins ("instruments common to the whole body") he indicated that the nerves all originate from the brain, the arteries from the heart and the veins from the liver. He commended Hippocrates for calling Nature just because the distribution of the blood vessels and nerves throughout the body is "in accordance with the value of each part." Because these structures passed safely to their destinations, he thought of Nature as being "not only just, but also skillful and wise" (May, 1968). Galen failed to differentiate nerves from tendons, and his inadequate and distorted account of the blood vessels prevented him from discovering the pulmonary circulation. Also, "his influence blocked the way for the discovery of the real circulation" (Sarton, 1954).

Of interest too, is Galen's description of the organs of reproduction and his views on fetal development. Unlike Aristotle, Galen thought of the ovaries as corresponding to the testes and of both producing sperms. He observed the uterus of animals and thought that the human uterus was also bicornuate.

Despite the many obvious errors in Galen's anatomical treatises, because of his reliance on animal dissection, one can still appreciate his meticulous description, verbose as it may seem. As an example, his account of the uterus and its peritoneal relationships is impressive.

Even though Galen assumed the uterus to consist of two horns, he correctly described the cervix, the vagina and the ovaries, as well as the arteries and veins. Both suspensory and round ligaments are mentioned and he considered these to be homologous to the cremasteric muscle which he had first described. More accurate, however, is his description of the peritoneal relationships with particular reference to the bladder.

COMPARATIVE ANATOMY

Galen meticulously dissected several species of animals, including the Barbary ape (*Macaca inuus*), the only European non-human primate. Galen carried out his anatomical studies during a period when dissection was looked upon as a pagan practice and not necessary for the treatment of patients. It was felt that what needed to be known about the human structure can be learned from the dissection of animals. Throughout his great work in anatomy he detailed the dissection of animals and without any hesitation assumed that his observations were the same for man (May, 1968).

In describing the cranial nerves, Galen stated that "for many surgeons do not know that in his work on the roots of the nerves Marinus has enumerated only those same roots which Herophilus specifies, but Marinus has concluded that there are seven pairs, whereas Herophilus says there are more than seven, regardless of the others. Whoever does not know this is, as the proverbial expression goes, like a seaman who navigates out of a book. Thus he reads the books on anatomy, but he omits inspecting with his own eyes in the animal body the several things about which he is reading." When describing the brain, Galen recommended "that the dissection is best made in apes, and among the apes in such a one as has a face rounded to the greatest extent possible amongst apes. For the apes with rounded faces are most like human beings" (Duckworth, 1962).

For studying the male reproductive organs, Galen gave specific dissecting instructions: "We say that in order for you to secure that the animal which you are dissecting resembles a man, you must take for that dissection an ape. But in order to achieve the effect of clarity in the appearance of such of those organs as are small and hard to see, then you must take a he-goat, a ram, bull or horse or male donkey for your dissection because that animal must necessarily possess a scrotum" (Duckworth, 1962).

> For studying the larynx, Galen recommended the use of the pig. For in all animals which have a larynx, the activity of the nerves and the muscles is one and the same, but the loathsomeness of the expression in vivisection is not the same for all animals. Because of that for my own part, as you know already, I illustrate such vivisections on the bodies of swine or of goats, without employing apes. But it is necessary that you should extend your studies and examine the larynx. This is constructed in the same way in the bodies of apes and men, construction which is shared by the other animals which have a voice. You must, then, dissect a dead man and an ape, and other animals furnished with a voice which have, besides the voice, the vocal apparatus, the larynx. For the animals which possess no voice have no larynx either. He who is not versed in anatomy thinks that in regard to the plan of the larynx great contrasts exist among the six Animal Classes, to which this our discussion refers. That is because neither the absolute dimensions nor the shapes of the parts of the larynx are precisely the same amongst all of them, a point which applies also to the number of the muscles which they have there. But as regards the activity of each one of the parts of the larynx, and the service which they perform, these are one and the same in all animals provided with a voice. That is because in the bodies of these animals the intention of the Creator was uniform with regard to the plan of the vocal apparatus, just as his intention was uniform also with regard to the plan of the respiratory organs in those animals provided with respiration. For the contrasts between these organs and the bodies of these animals consists solely in their absolute dimensions and their shapes (Duckworth, 1962).

Galen's accomplishment in carrying out actual dissections in animals and his detailed descriptions of those structures he observed, continuing in the tradition of Aristotle, would clearly make him one of the earliest comparative anatomists.

PHYSIOLOGICAL SYSTEM OF PNEUMAS

From simple experiments he carried out, Galen demonstrated that arteries contained blood and not air as was previously thought. But his dogmatic adherence to the pneumatic theory (Fig. 27) prevented him from understanding more about the flow of blood

(Singer, 1957; Siegel, 1968). Essentially, three types of special pneuma or spirit were thought to exist which dominate the liver (natural spirit), the heart (vital spirit) and the brain (animal spirit).

Galen thought that pneuma, which was the life force, was taken into the body during breathing. Pneuma entered the lungs via the trachea and from the lungs it passed into the left ventricle through the "vein-like artery" (pulmonary vein). He demonstrated the presence of blood in the left ventricle, but believed that it originated in

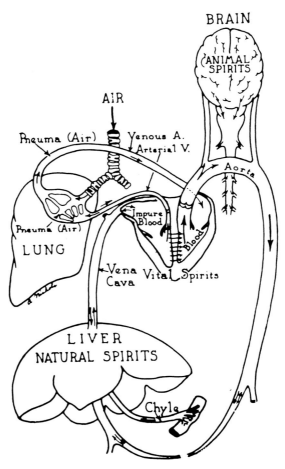

Figure 27. Galen's conception of the vascular and respiratory systems — the flow of *pneuma*. (With permission, from Major 1954.)

the liver from chyle, derived from the intestinal tract, where it became imbued with "natural spirit."

Galen believed that this activated blood ebbed and flowed through the venous system, including through the artery-like vein (pulmonary artery), to the lungs where impurities were removed. He held the view that the venous blood in the right ventricle flowed back into the general venous system but some of it was filtered through tiny pores in the interventricular septum to the left ventricle. Here it came into contact with the pneuma and this gave rise to the "vital spirit" which was then distributed by the arteries. When the vital spirit reached the brain through its arteries, it was changed into the animal or psychical spirit. This vital spirit was conceived as being distributed through canals in the nerves.

The natural spirit in the liver was thought to control the functions of nutrition, growth and reproduction whereas the vital spirit of the heart regulated vital functions by transporting heat and life throughout the arterial system; the vital spirit of the brain regulated the brain itself. This pneumatic theory, which originated during the Hippocratic period, was based on three dominant and regulating forces or spirits in the body on the basis of which all functions can be explained. Philosophical in its formulation and completely lacking any scientific basis, Galen's physiological system of pneumas survived up to the 17th century when Harvey announced his discovery of the circulatory system.

EXPERIMENTAL STUDIES

Notwithstanding Galen's highly fanciful concept of bodily functions, based on the theory of the three pneumas (Singer, 1957; Siegel, 1968; May, 1968), he carried out a large number of anatomical and physiological experiments which were of a practical nature and had clinical implications. He performed experiments on the lungs in order to understand more about respiration, studied digestion by feeding pigs with different diets and then opening their stomachs, proved that the arteries contained blood and not air by placing two ligatures on an exposed segment of artery which he then slit open.

Galen accurately described the consequences of spinal cord damage at different levels. He observed loss of sensation and paralysis of all muscles supplied by nerves originating from the spinal cord following complete resection below that level. In contrast, he found that hemisection of the spinal cord resulted in paralysis of the muscles only on the side with the lesion. Furthermore, longitudinal medial incision of the cord did not lead to paralysis.

He showed that in addition to the diaphragm other muscles were involved in respiration. His detailed description of the origin and course of the phrenic nerve and the accidental discovery of the recurrent laryngeal nerve, which led him to understand something of voice production in the larynx, is most impressive.

SCIENTIFIC ACHIEVEMENT

In retrospect, Galen deserves greater recognition for his work than he has received in the past, and as mentioned in the introduction to Duckworth's translation of the later books on *Anatomical Procedures*, the avowed aim was to "focus the attention of scholars and scientists on one of Galen's greatest works, and to attempt to remedy the accidental, but disastrous dichotomy from which it has suffered for so many centuries." The hope was also expressed that "modern anatomists, with their interest in comparative as well as in human anatomy, and with their experimental approach towards the elucidation of anatomical problems, are likely to be more sympathetic towards Galen than have been their predecessors during the last 400 years. The eclipse of Galen's reputation at the Renaissance was perhaps inevitable. But if in recent centuries he has been remembered chiefly for his errors, it is possible now to put these into perspective against a background of solid scientific achievement" (Duckworth, 1962).

For a fitting tribute to Galen one should perhaps refer to the commentary of Hunain ibn Ishaq of Baghdad (A.D. 809-873), the great medical scholar, linguist and Galen translator, who upon completing his translation of the fifteenth book commented that "this excellent, outstanding work which is one of the compositions of a man who performed marvellously, and revealed extraordinary things, the master of the earlier surgeons, and the lord of the more recent savants, whose efforts in the practice of medicine have been unequalled by any of the prominent since the days of the learned and great Hippocrates — I mean Galen. May God Almighty be merciful to him!" (Duckworth, 1956).

Galen is often described as the Prince of Physicians, second only to Hippocrates, the Father of Medicine. He was the last of the great medical scholars of antiquity, and his death, 200 years after the Christian era began, coincided with the progressive decline of Greek science. Through his many works, Galen has left for posterity the medical accomplishments of an era that would have been all but forgotten.

CHAPTER EIGHT

THE EARLY MIDDLE AGES

ROMAN AND BYZANTINE EMPIRES

WITH the decline of Greek science and culture and the fall of the Roman Empire dawned a new period that was not seminal ground for new ideas. What constitutes the Middle Ages is still a matter of debate among historians. It probably began long before the Roman Empire ceased to exist in A.D. 476 and came to a close during the 15th century with the fall of the Byzantine Empire in 1453.

Because of its conformity with the philosophy of the governing stoics and the teachings of the Christian Church, Galen's work prevailed for more than 1300 years. It was believed that everything one needed to know about the structure of the human body can be found in Galen's anatomical treatises which became established as the ultimate authoritative source.

Galen's death heralded a long era with a predictable outcome. Medicine, and the study of human anatomy in particular, languished in passive moribundity only to reach a climactic end in 1543 with the publication of Vesalius' *De Corporis Humani Fabrica*, a year which also coincided with the publication of Copernicus' monumental work *De Revolutionibus Orbium Coelestium*. These two works revolutionized the thinking of man with respect to the structure of the human body and the universe, respectively. Indeed, modern medicine is considered to begin from the time of Vesalius. It is not surprising therefore for medical writers to consider the Middle Ages as the period from the death of Galen in A.D. 200 to the publication of Vesalius' work in 1543.

By A.D. 400, Christianity had spread throughout the Roman Empire and the power and influence of the church grew to the extent of becoming tyrannical, so much so that there was no interest in learning and scholarly pursuits. The work of the church took precedence with its emphasis on the saving of souls and the reinforcement of theological dogma.

Despite the efforts of the great Roman Emperors, Diocletian, who ruled from 284 to 305 A.D. and Constantine I who ruled from 312 to 337 A.D., the downfall of the Roman Empire could not have been prevented because of political problems relating to the choosing of successors, high taxes and inflation, and innumerable attacks by the Germanic tribes from Northern and Central Europe which began in 378 A.D.

Following the fall of the Roman Empire in 476 A.D., the Germanic conquerors began to establish kingdoms which eventually proved to be unmanageable. In contrast, the Byzantine Empire, with its capital in Constantinople, flourished. The emperor, Constantine, had moved the capital of the Roman Empire from Rome to Constantinople in 330 A.D. The city itself was the most magnificent, largest and affluent in the whole of Europe. Its citizens were cultured, educated and creative. Within this advanced civilization, Greek culture, as well as Roman moral values and political systems, was nurtured for almost a thousand years.

RISE OF ISLAM AND ARABIAN MEDICINE

Political and military powers in the hands of the nobility, lack of a central government, the absence of democracy and the decline of trade plunged western Europe into a period of darkness that lasted until about 1000 A.D. Western civilization owes an enormous debt to the Arabs for preserving the great achievements of the Greek scholars, particularly in medicine and in the sciences (Moir, 1831; Brown, 1921; Campbell, 1926; Ashoor, 1984; Hayek, 1984).

The Prophet Mohammed, who was born in 570 A.D. in the holy city of Mecca, founded the religion of Islam (meaning submitting oneself to God) based on the revelations he received while meditating in a cave when he was 40 years old. Leiser (1983) in his excellent study of medical education in Muslim countries during the Islamic era indicated that the Prophet's remarks (Hadiths) in regards to medical matters, comprising the *tibbi al-nabī*, probably influenced folk medicine but were of lesser importance in the formal training of physicians. As far as human anatomy is concerned, some of the Hadiths and citations in the Qu'ran with respect to intrauterine development are outstandingly remarkable. The following sequences have been described (see Moore, 1982; Albar, 1983):

1. *Nutfa* (or drop stage)—"We created Man from mixtures of germinal drop" and "From a *Nutfa* He hath created him, immediately programmed him."
2. *Alaca* (referring to early implanting embryo)—"Then We made the sperm into a leech-like clot" (Sura XXIII-14).
3. *Mudghda* (somite stage)—"then of that leech-like clot made a chewed-like substance" (Sura XXIII-14), which is mentioned as being "partly differentiated and partly undifferentiated."
4. "Then We made out of that chewed like substance bones" and "clothed the bones with flesh: then We developed out of it another creature." These are obvious references to the formation of the skeletal and muscular systems, as well as the transition of the embryo to a new stage with more definitive and recognizable features.
5. Finally, with reference to differentiation between male and female fetuses (external sexual characteristics) is the Hadith "When forty two nights have passed over the sperm-drop, Allah sends an angel to it, who shapes it and makes it ears, eyes, skin, flesh, and bones. Then he says, "O Lord; is it a male or female? And your Lord decides what he wishes and the angel records it."

United by Islam and guided by the Qu'ran, the Arabs were eager to spread their new religion, and their armies with the conquest of Byzantine Asia, Persia, Egypt, North Africa and Spain now dominated an empire larger than that of the Romans. Having encountered here the dispersed Greek physicians and philosophers, as well as their great libraries, the Arab scholars began to study the works of the conquered people, particularly in medicine and the physical sciences, and encouraged their translation into Arabic. These physicians-translators were chiefly Syrians, descendants of Nestorian Christians, but also included Persians. They pursued their task with great passion and vigor as a result of which many important and major works of Greek scholarship (Hippocrates, Archigenes, Dioscorides, Rufus of Ephesus, Galen, Oribasius, Philagrios, and Paul of Aegina) were lodged in the flourishing libraries of the Arab empire, particularly at the "House of Science" in Baghdad, its capital (Gordon, 1959; Von Grunebaum, 1963; Klein-Franke, 1982; Ashoor, 1983, 1984).

The original Greek words in anatomy were translated and would have been lost to western civilization were it not for their rediscovery during the 13th century when they were translated back into Latin. Some of the Arab doctors might have been taught Galen's anatomy by Paul of Aegina who was in Alexandria in A.D. 640 at the time of the Arab Conquest.

The golden age of Arabian medicine lasted until about the middle of the 12th century. Between 800 and 1100 A.D., Arabic was the most widely accepted language for communication among medical scholars (Leiser, 1983). All major works were already translated and widely disseminated in the flourishing libraries of the Islamic empire. The library in Cordoba alone housed more than six hundred thousand volumes and a catalogue of forty-four volumes. These were largely commentaries based on the translations of Hippocrates and Galen's work.

The intellectual efforts of Arab scholars were not limited to the translation and assimilation of the works of the Greeks but independently they have made spectacular contributions in mathematics, astronomy, chemistry, physics, botany and pharmacy (Moir, 1831). They knew of the use of many herbs and chemical preparations for the treatment of diseases. Indeed, the first schools of pharmacy and dispensaries can be traced to this period.

Arabian medicine has produced many great physicians whose works profoundly influenced medical thinking when they were later translated from the Arabic into Latin (Browne, 1921; Campbell, 1926; Gordon, 1959; Leiser, 1983; Ashoor, 1984; Hayek, 1984; Shanks and Al-Kalai, 1984). The most famous of these were Rhazes, Haly Abbas (Fig. 28), Avicenna, Avenzoar (1113-1161 A.D.), Averroes (1166-1198 A.D.) and Hunain ibn Ishaq.

Rhazes (860-932 A.D.) was the author of more than 200 books and to him we owe the first description of measles which he clearly distinguished from small pox. Two works have survived: *Al Hawi* or

Cum priuilegio Pontificis maximi & comis deci
mi:z Francisci christianissimi Francorum regis.

·YSAAC·

HALY ABBAS

COSTANTINVS MO.

Omnia opera Ysaac in hoc volumine con
tenta: cum quibusdam aliis opusculis·

Figure 28. The three founders of early mediaeval anatomy from the title-page of the Opera Ysaac, 1515. Haly Abbas is shown on the left; Constantinus Africanus is on the right. (Corner 1927, with kind permission from the Carnegie Institution of Washington.)

Continens, an encyclopedic medical compilation of 24 volumes and a shorter *Almansor* which were essentially composed of extracts from hundreds of earlier writers. They were translated in the middle of the 12th century by Gerard of Cremona. The *Al Maleki*, a manual of medical and surgical practices, was compiled by Haly Abbas (?-994). This was the standard work for doctors training in European schools until Avicenna's (980-1037) *Canon of Medicine* appeared in A.D. 1000. It contained a fair amount of anatomy which was borrowed from the writings of Aristotle, Hippocrates, and Galen in particular (Gruner, 1930). In addition to translating as many as 129 of Galen's books and a variety of Greek medical and scientific works into Syriac and Arabic, Hunain (A.D. 803-873) also compiled several medical treatises including a survey of medical knowledge and a book dealing specifically with eye diseases (Bergsträßer, 1925;

Meyerhoff, 1926, 1928). With the translations, Hunain was assisted by his son Ishaq and nephew Hubaysh al-As'am.

ARAB SCHOLARS AND ANATOMY

In order to prevent quackery and charlatanism, examinations were proposed for testing medical knowledge. One of the required subjects was anatomy. Arab physicians pursued their work with zeal and made many meticulous observations of diseases (Browne, 1921). Perhaps, given the opportunity they would have made more advances in human anatomy were it not for their religion, which like so many others, forbade human dissection (Leiser, 1983).

Ali ibn Rabben at-Tabari, a physician of the 9th century, dealt with embryology, as well as other matters, in his work *Paradise of Wisdom*. In his *Kitab al-Mujiz*, Ibn Nafees (1210-1288 A.D.) asserted that the interventricular septum of the heart had no openings and that blood flowed from the heart to the lungs and from the lungs to the left side of the heart by the pulmonary veins. Thus, for the first time, the pulmonary circulation is mentioned (Klein-Franke, 1982; Ghalioungui, 1982). This discovery has been traditionally attributed to Michael Servetus because he also provided an accurate description of the lesser circulation in his book *Christianismi Restitutio* which was published in 1553 (Lambert, 1936).

In the pre-Islamic era, various parts of the human body were known to be important but as to their shapes and functions the description was more poetic than factual and scientific (Hyrtl, 1879; Browne, 1921; Campbell, 1926). Within the heart were to be found courage and passion, anger came from the liver, and from the lungs and spleen emanated fear and laughter, respectively.

Between the 7th and 9th century Islamic medicine derived much from the earlier Greek translations of these texts. A new phase gradually emerged under the influence of medical knowledge and philosophy from the Greeks, Syrians, Persians and Indians. It was the period during which the most famous of Arab physicians lived. It was also a period of original thinking and major contributions to medical sciences, including human anatomy.

Avicenna in his Canon (Gruner, 1930) presented a classification of the organs and their functions. The eyes are described, including six muscles which are responsible for movements of the eyeball and paired optic nerves which crossed. He commented on the functions of the lacrimal glands and the iris, and even speculated that the images from both eyes superimposed on each other. The most original contribution to the descriptive and functional anatomy of the eye was to be made later by an Arab mathematician and physicist, Al-Hazen or Ibn-al-Haytham (965-1040 A.D.) in his work *Kitab al-Menazir*. He deviated from humoral and teleological traditions in order to formulate a theory of vision based on his understanding of the eye as an optical system (Russel, 1982).

Haly Abbas described the heart and circulation in remarkable detail. The two ventricles are mentioned, as well as their openings. The valves and their functions were recognized. It was even pointed out that the two ventricles contracted at the same time, but the left more forceful sending blood throughout the body. It was Haly Abbas who also described the liver as having two or three lobes in contrast to the previous description of five lobes by Hippocrates and Galen.

Contrary to Galen's description of the mandible as consisting of two parts joined in the middle, Abdu-l-Latif (1162-1231) of Baghdad described it as a single bone after examining a large number of skeletons during a visit to Egypt (Karim, 1982).

According to Mettler (1947) the hot and dry climate of the countries in the Arabian peninsula did not favor the preservation of the body for the purpose of dissection. Lassek (1958) stated that "The Quran was the source and authority for all knowledge; anatomical dissection was forbidden; even drawings and sculpture of the human body were not permitted. It was held that the soul suffered if the body was incised before or after death. They knew Galen's anatomy and not much more; this lasted until modern times. As late as 1830, the Arabs had only a pristine knowledge of the subject."

Evidence is now accumulating that such was not the case and that early Arab physicians, including ibn Nafees who discovered the circulation in the lung, practiced dissection (Nasr, 1976; Qatagya, 1982; Ahmed, 1982; Uddin, 1982). Even among contemporary Arab doctors and scholars this problem has provoked heated discussions (El-Gindy and Hassan, 1982) and the debate will undoubtedly continue. In this regard, attention should be directed to the anatomical illustrations preserved in several mediaeval treatises, all perhaps copies derived from Islamic sources (O'Neill, 1969, 1977, 1982).

It is now believed that the original *Fünfbilderserie*, consisting of five anatomical drawings found in two Bavarian monastic manuscripts and reported in 1907 by Sudhoff, can be traced to an original manuscript with nine sets of drawings. The complete drawings, as well as the sequence according to the anatomy of Galen, have been preserved as part of a medical codex in the library of Gonville and Caius College in Cambridge, England.

As pointed out by O'Neill (1982) the famous Ashmolean anatomical drawings (Figs. 29-32) of the Bodleian Library in Oxford, England, are essentially the same as those present in Cambridge, except for a drawing of the heart emerging from the sheath-like lungs. In addition, the normal sequence of the drawings has been disrupted. The Cambridge drawings reveal Arabic influences which are not evident in the manuscript that was studied by Sudhoff. According to O'Neill (1982) the Gonville and Caius manuscript present "the most complete and graphically intact Latin copy of the treatise known."

O'Neill (1982) suggested that "an illustrated Arabic anatomical manual found its way from Spain into an English Benedictine monastery" and thereby reached European scholars. Fanciful and interesting as they are, these anatomical illustrations are among the earliest known. Galenic in concept, these crude illustrations reveal no practical knowledge of the inside of the human body (Figs. 33 and 34).

EUROPEAN UNIVERSITIES

Europe was experiencing a gradual transformation because of a change in societal values, improvement in trade, expansion of the cities, and the establishment of strong governments. With the fall of the Roman Empire the authority of the Pope diminished and religion became more unified. The discovery of America in 1492 by Christopher Columbus and of the art of printing with moveable type by Gutenberg in 1454 were all momentous achievements. There was renewed interest in every aspect of human endeavor and the time for exploration of new ideas and discoveries was eminent. It was also the period that gave birth to the great universities and medical schools (Gordon, 1959) which fostered a renewed interest in the anatomy of the human body. Prior to this period anatomy was considered to be a

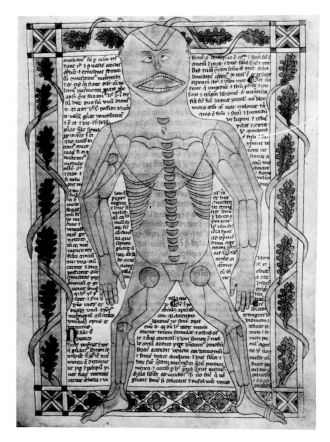

Figure 29. Crude drawing of the skeleton from a 13th century English manuscript. (Ms. Ashmole 399, folio 20; Bodleian Library, Oxford.)

pagan science and looked upon with contempt. Such an attitude was a reflection of the psychology of the times.

There was a great preoccupation with religion, the occult, mysticism and astrology. The human body and its functions, as well as man's fate, were thought to be under the magical influence of the planets. Because every part of the body was governed by one of the heavenly bodies, then it followed that diseases could be accounted for by the movements of the planets. The zodiacal man became popular and accounts for the many anatomical drawings, decorated with

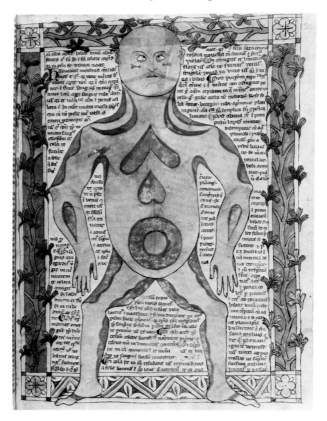

Figure 30. The muscular system; an imaginary representation from a 13th century English manuscript. (Ms. Ashmole 399, folio 22; Bodleian Library, Oxford.) Peculiar arrangement of muscles, probably an attempt to relate various bodily movements to specific muscle masses.

astrological signs (Figs. 35 & 36), which have emerged from this period (Major, 1954; Lassek, 1958).

The renewed interest in human anatomy was intimately linked with the emergence of the universities, but it still had to be legally approved by both church and government. In Italy alone, fifteen universities were established between A.D. 1200 and 1350. The first center of learning to be established was the University of Salerno, but as far as human dissection was concerned, it would appear that this occurred at the University of Bologna. The reason was not for

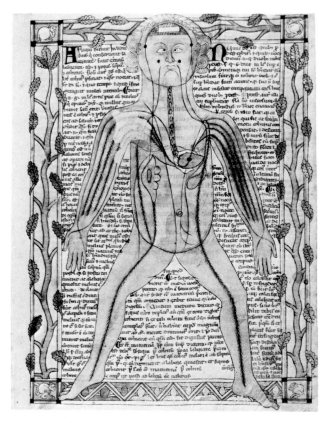

Figure 31. Drawing of the arterial system from a 13th century English manuscript. (Ms. Ashmole 399, folio 19; Bodleian Library, Oxford.) The black spot in the heart was conceived as the site of origin of the arteries. The anastomosis of the two arteries in the head probably represented the *rete mirabile*. Observe the characteristic squatting or "frog-like" posture of these mediaeval anatomical figures.

the teaching of anatomy but for ascertaining the cause of death in a suspected case of poisoning.

The University of Bologna began for the teaching of law and the medical faculty was established in 1156. It is stated that by 1320 there were at least 15,000 students there. One of the first anatomists to have taught at Bologna was Taddeo Alderotti, who carried out human dissections in the teaching of anatomy (Siraisi, 1981), according to the records of his disciples Bartolomeo da Varignana, Henri de Mondeville and Mondino de Liuzzi. It is also recorded that post-

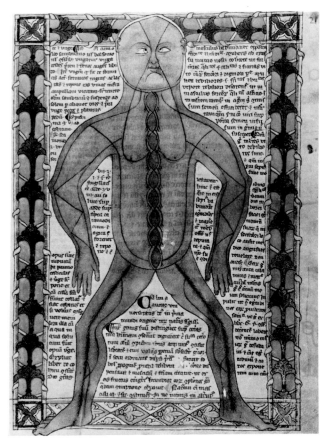

Figure 32. Speculative concept of the nervous system from a 13th century English manuscript (Ms. Ashmole 399, folio 21; Bodleian Library, Oxford).

mortem autopsies were carried out on the body of a nobleman, Azzolino, who died in 1302, under the direction of Bartolomeo da Varignana (Simili, 1951; Siraisi, 1981), a student of Taddeo and also by the surgeon William of Saliceto, a contemporary of Taddeo. The authorization to carry out human dissections is attributed to Frederick II (1194-1250), Emperor of the Holy Roman Empire and King of Jerusalem.

As many as eighty universities were established in Europe before the end of the Middle Ages, but nowhere was the study of human anatomy pursued with such vigor as in the universities of Italy. The

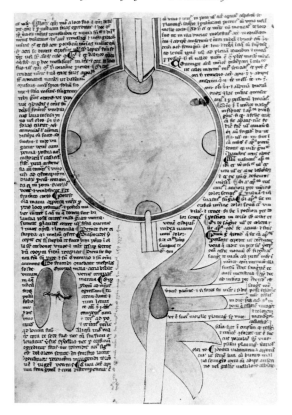

Figure 33. Drawings of a spherical stomach, paired kidneys, and a five-lobed liver from the Ashmole 399 (folio 23) 13th century manuscript (Bodleian Library, Oxford). These illustrations were copied from other sources and bear little or no relationship to the actual organs.

most celebrated of them was the University of Bologna.

At the University of Salerno, the earliest known university in Europe, many anatomical treatises were written and translated (Harrington, 1920). It was here that Constantinus Africanus re-translated the works of the Greek and Arab scholars into Latin (Steinschneider, 1866) and the younger Cophon wrote his famous *De Anatomia Porci* in the 12th century (Corner, 1927) which served as a standard work of reference for the learning of anatomy. Dissections were carried out in the pig, monkey and bear. Because the edict of Frederick II in A.D. 1240 required all surgeons to study anatomy

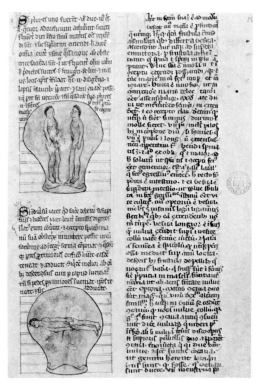

Figure 34. Position of the infant in the womb as depicted in this 13th century English manuscript. (Ashmole 399, folio 15; Bodleian Library, Oxford). Above are twins, male and female; below, the child lies transversely with attached umbilical cord.

for a year and to demonstrate their competence in this subject, permission was granted for one human dissection every five years.

Corner (1927) viewed Salerno as the medical school "through which European medicine was awakened after the Dark Ages." In his monograph of some of the anatomical works of the earlier Middle Ages, one is informed that "it is now clear that the School of Salerno was not a miraculous relic of ancient Greece, but a new structure upon a foundation slowly laid down toward the end of the Dark Ages, with the aid of influences from across the Mediterranean." We are informed that before the year 1050 medicine was practiced chiefly in the Benedictine Monasteries, as was the pursuit of other

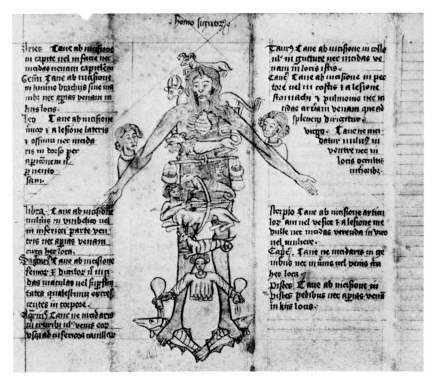

Figure 35. Astrological Man. Wellcome Ms. 40. Calendarium 1463. (Wellcome Institute Library, London.) In the Middle Ages and during the Renaissance, astrology was taught in several European universities and many physicians were also astrologers. It was believed that the planets ruled different parts of the human body. Thus, Jupiter — lungs, liver and feet; Mars — left ear and genitalia; Sun, right side of heart; Venus — neck and abdomen; Mercury — arms, hands, shoulders and hips; Moon — left half of body and stomach.

branches of learning. Between the 8th and 14th centuries many manuscripts were composed in Latin and of these more than 600, including fourteen medical volumes, have survived. Almost nothing was mentioned about the structure of the human body. However, the best known work of mediaeval medicine, was the poem *Regimen Sanitatis Salernitanum*. It emanated from Salerno and spread throughout the civilized world even before the advent of printing. By 1846, more than 240 editions of the poem were printed and 100 copies of the manuscript were in existence (Packard, 1920).

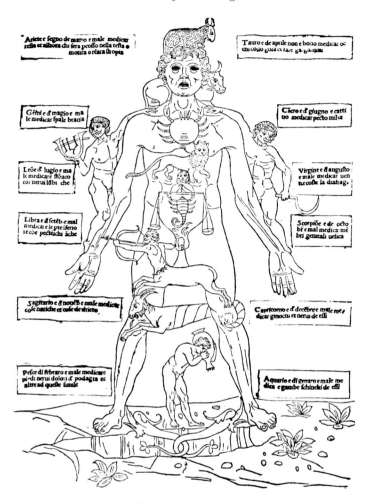

Figure 36. "Zodiac man" from Ketham's *Fasciculus Medicinae* (1522). The astrological symbols are shown in relationship to the different parts of the body (Instituti Ortopedici Rizzoli, Bologna). The human body as the microcosm or parvus mundus was viewed and interpreted in relationship to the universe or macrocosm. The 12 signs (constellations) of the Zodiac were represented as follows: Aries, the ram; Taurus, the bull; Gemini, twins; Cancer, the crab; Leo, the lion; Virgo, the virgin; Libra, the balance; Scorpio, the scorpion; Sagittarius, the archer; Capricorn, the goat; Aquarius, the water-bearer; and Pisces, the fish.

Constantinus Africanus (about 1020-1087) arrived in Salerno about the year 1070. He was born in Carthage and for thirty-nine years travelled throughout the Orient, India, Babylon, Ethiopia and Egypt. He knew many languages, including Greek, Latin and Arabic, and translated between twelve to fifteen medical texts from Arabic into Latin. These included the works of Hippocrates, Galen, and that of Haly Abbas (Sudhoff, 1930; Lehman, 1931). The work of Haly Abbas was entitled *Al Maleki* (The Royal Book) which attempted to summarize the entire system of classical medicine. It was written by the author during the latter part of the 10th century and consisted of two parts, *Pars Theoretica* and *Pars Practica*. Each part was subdivided into ten books and the two books (Books 2 and 3) of the *Pars Practica* in Constantine's translation of the work, which he entitled the *Pantegni* or *The Whole Art*, presents a systematic account of the human body based on the teaching of Galen. In 1539, a collected edition of his books was published in Basle (Packard, 1920).

Accused of practicing sorcery, Constantinus Africanus fled to the silent cloisters of Monte Casino in 1076 where he died eleven years later. According to Sudhoff and Corner, Constantine "must be held among the founders of modern medicine and indeed of all modern biology. He gave the West a great mass of important classical learning, in readable Latin, at a time when everything was ripe for growth. For a hundred years, all western medicine grew out of these books, and when the 13th century brought new translators who were able to comprehend and to translate Aristotle and Avicenna, their seed fell upon a ground ploughed by Constantine."

Salerno was the most prominent medical school of the earlier Middle Ages. The teaching of anatomy was based on the translated works of the Greek, the Graeco-Roman and Arabic scholars. The pig was dissected in order to study its internal organs. They were believed to be similar to those in man. Corner (1927) described the sources and content of five anatomical manuscripts which have emanated during this period.

Legend states that the medical school in Salerno was open to female students. The best known is Tortula, a noble woman, who apparently also taught at Salerno and is credited as the author of two medical books, *De Mulierum Passionibus* and another on cosmetic hygiene.

For the practice of medicine standards were already established in 1140 by Roger II of Sicily who decreed that candidates must demonstrate their competence and knowledge by passing appropriate examinations, and in 1224 Emperor Frederick II, his grandson, made public examination of all candidates by the teachers in Salerno a requirement for practicing medicine. In order to be eligible for the examination, the candidate had to prove the legitimacy of his birth, to have attained the age of twenty-one years and must have studied medicine for seven years. The works of Hippocrates, Galen, Avicenna and Aristotle were examined and the title of Magister was conferred on successful candidates. To practice as a surgeon, it was necessary to study anatomy at Salerno or Naples and to pass a rigorous examination (Packard, 1920). The influence of the University of Salerno, as well as other factors, contributed to the development of similar centers of higher learning in other parts of Italy, France, Germany and England.

The University of Montpellier apparently had *magistri physici*, medical teachers, as early as A.D. 1000 which would make it the oldest university in Europe. It was founded by Jewish physicians from Spain and like Salerno, which it rivalled, was the center for the dissemination of Greek medicine to Western Europe. The Bishop appointed the teachers and in A.D. 1230 the passing of an examination conducted by two of the masters was a requirement for practicing medicine. Here, the learned Catalan Arnold of Villa Nova (1235-1311) translated Avicenna's book on the heart and wrote other medical treatises. He was an independent thinker, highly regarded, and knew many languages. As physician to several popes and the Kings of Arragon and Sicily he was a man of considerable influence (Packard, 1920).

After 1314, dissections were carried out in Montpellier and it was opened to the public on the payment of a fee. The Duke of Anjou ordered in 1376 that the body of an executed criminal should be made available for dissection annually. The famous surgeon, Guy de Chauliac (1300-68) studied in Montpellier and Henri de Mondeville (1260-1320), another surgeon who encouraged the study of anatomy (Pagel, 1889; MacKinney, 1962), was a professor there.

In mediaeval France, medicine was taught only at Montpellier and later at the University of Paris which was also renowned for theology and philosophy. The University was founded during the latter part of the 12th century and attracted students from all over

Europe. The teaching of medicine probably began in 1210 when Gilles de Corbeil of Salernum was appointed physician to King Philip.

The University of Oxford began in 1167 and the University of Cambridge in 1217, both patterned after the University of Paris where the courses in medicine were the same and at the beginning did not include instruction in surgery.

CHAPTER NINE

MONDINO DE LUZZI

RESTORER OF ANATOMY

THE practical aspects of medicine, in particular surgery, gradually acquired increasing importance which warranted a greater familiarity with the structure of the human body. To do so required human dissection and for this we must turn our attention once again to the University of Bologna where the celebrated anatomist, Mondino de Luzzi (1276-1326), taught and carried out for the first time again, after more than 1700 years, one of the first dissections of a human body (Pilcher, 1911; Schipperges, 1959; Kudlien, 1964; Nannini, 1967). From the available records, this was done on an executed female in 1315 for the benefit of medical students, as well as the public.

The Professor sat on an elevated chair and read from the work of Galen to an assistant, who did the actual dissection, and a demonstrator who pointed out the various parts (Fig. 37). Because the cadaver was unembalmed, the dissection proceeded rapidly in four sessions which covered the digestive, respiratory, circulatory, and muscular systems.

Mondino compiled one of the earliest anatomy textbooks, *De Omnibus Humani Corporis Interioribus Membris Anathomia*, which contained copious instructions for dissecting (Ketham, 1491; Dryander, 1541; Shipperges, 1959; Kudlien, 1964). The book, although completed in 1316, was first published in 1487 at Padua and six years later in Leipzig (Melerstat, 1493; Fig. 38). It was a small volume of

Figure 37. Anatomical dissection scene from Ketham's edition of Mondino's *Anathomia* (Venice, 1500). Mondino is shown seated on the elevated chair reading from a book and presiding over the dissection. The assistant is about to make an incision. (Instituti Ortopedici Rizzoli, Bologna.)

only forty-four pages, without illustrations, and essentially reiterated Galen's anatomy. Mondino's *Anathomia* was extremely popular and went through more than forty editions (Fig. 39) because it embodied a unique blend of imagination and innovation. The first anatomical book with illustrations was Ketham's *Fasciculus Medicinae* (Fig. 40) which was published in Venice in 1491. Subsequent editions showed a picture of Mondino seated on an elevated chair and supervising a dissection (Fig. 37). At the beginning Mondino dissected himself but later this task was passed on to his assistants. In this context, it should be noted tht the first printed picture of an anatomical dissection is to be found in the French translation of Bartho-

Figure 38. Title-page of the 1493 edition of the Anathomia published in Leipzig. The man in full robes and cap sitting on the chair is undoubtedly the professor. He is reading from the opened book in his left hand and directing the young bareheaded assistant. The intestines are shown.

lomaeus Anglicus' encyclopedia, *De Proprietatibus Rerum*, published in 1482 in Lyon (Fig. 41). It was Mondino's anatomical demonstration on the human cadaver that sparked a revolution in the teaching of anatomy, and for this reason Mondino is often referred to as the "restorer of Anatomy."

Mondino was born in Bologna where he studied medicine and probably acquired an interest in anatomy by observing his teacher, Thaddeus, performing postmortem examinations. He graduated in

Figure 39. An anatomical demonstration in Rostock at the beginning of the 16th century. This wood-cut illustration appeared in the Rostock edition (1514) of the *Anatomia Mundini* by Nicolaus Thurius Marschalk (Universitätsbibliothek Rostock; from Wischhusen and Schumacher, 1970). The first anatomical dissection and demonstration in Rostock took place in 1513; for some other European cities the following years have been cited (see Wischhusen and Schumacher, 1970): Bologna (1302), Montpellier (1315), Padua (1341), Prague (1348), Venice (1368), Florence (1388), Vienna (1404), Paris (1478), Cologne (1479), Tübingen (1482), Leipzig (c. 1500), Edinburgh (1505), Strassburg (1517), Basel (1531), Marburg/L (1535), Oxford (1549), Salamanca (c. 1550), Lausanne (c. 1550), Zürich (c. 1550), Amsterdam (1555), Cambridge (1557), London (1564), Bern (1571).

1290 and became a Professor of Anatomy at the University in 1306 where he remained for the next twenty years until his death.

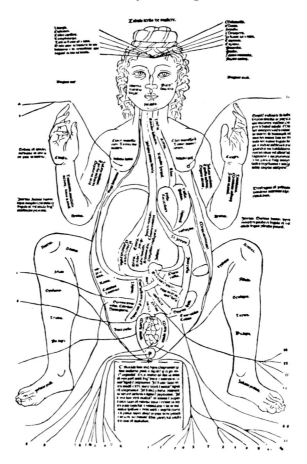

Figure 40. Situs figure from Ketham's *Fasciculus Medicinae*, Venice, Johann et Gregorio 1491 (From Major, 1954).

MONDINO'S ANATHOMIA

The *Anathomia* can rightly be considered the first modern work that dealt exclusively with anatomy (Kudlien, 1964; Shipperges, 1959). Even though it was inspired by translations of the Arab scholars from the works of the Greeks, the arrangement of the book followed the dissection procedure adapted by Mondino.

Because there was no way of preserving the cadaver, the dissection was carried out over four successive days and sometimes

Figure 41. First printed picture of an anatomical dissection (From the French translation of the *De Proprietatibus Rerum of Bartholomaeus Anglicus*, Lyon 1482; from Major, 1954). See also Figure 42.

included the nights. The alimentary canal was dissected first on account of putrefaction and this was followed by the spleen, liver, and great vessels of the abdomen. After dissecting the reproductive organs, the thorax was then opened. This was then followed by the dissection of the head and finally the trunk and extremities.

The anatomical nomenclature was inconsistent and largely derived from Arabic, including such terms as *mirach* (abdominal muscles), *siphach* (peritoneum) and *meri* (esophagus). It was the language that kept anatomy alive for many centuries and even to this day many Arabic names have survived as part of our modern terminology.

The work is strictly descriptive, often elementary and inaccurate, but an attempt was made to correlate structures of organs with their functions. There is reference to dissections Mondino had carried out in 1315 on two female cadavers in order to compare the size of the uterus in a virgin and in someone who had given birth.

Figure 42. Autopsy scene from an early 14th century manuscript of English origin (Ms. Ashmole 399, folio 34; Bodleian Library, Oxford). According to Singer (1956) it is the earliest known representation of a dissection. The physician, accompanied by a monk, seems to be giving instructions to the person dissecting. The cadaver is female. The operator has a knife in his left hand and in his right hand holds the liver. The kidneys, lungs, stomach, intestines, and heart are shown scattered. For other interpretations of this interesting miniature, see MacKinney 1965.

Although Mondino dissected the human body and probably made many worthwhile observations, he never questioned the authoritative writings of Galen and Avicenna, even when the findings were contradictory. He described the stomach as being spherical and the liver as being made up of five lobes. The gall bladder and the spleen together with their presumed humors, yellow bile (choler) and black bile (melancholia), respectively, as well as the many imaginary channels leading from them into the cardiac region of the stomach, were described.

Mondino identified one of the ducts of the pancreas but his description of this organ was very poor. He knew of the cecum but not of the vermiform appendix. The different origins of the spermatic vessels on the two sides were noted, but he described seven cells within the uterus, three ventricles were described for the heart as well as for the brain, and like Galen he identified seven pairs of cranial nerves and regarded the brain as the seat of intelligence and sensation.

Mondino dried some of his specimens in the sun in order to study the tendons and ligaments. He was able to trace the course of nerves by studying macerated specimens, but refrained from studying the bones because "owing to the sin involved in this I am accustomed to pass them by" even though he instructed, when describing the bones of the skull, that "you will see them better if you boil them." This was in accordance with the papal edict of Pope Boniface XIII, issued in 1300, which was directed against the practice of boiling the remains of those who died far away from their homes, in order to make the transportation of their bones easier for burial in their city of origin. Such was the custom among the Crusaders and the papal law aimed at discouraging it. This law was often interpreted by church authorities and many anatomists as a prohibition of human dissection. It was therefore even more remarkable that the practice of dissection should be revived within just a few decades.

Mondino's method of teaching anatomy had wide appeal and eventually it became quite common throughout Europe: at the University of Padua in 1341 by Gentile da Foligna, Venice in 1368, Florence in 1376, Paris and Vienna in 1404, Siena in 1427, Genoa in 1482 and Pisa in 1501.

A gall-stone was discovered in the first cadaver that was dissected at the University of Padua and similarly a stone was found in the bladder of the Bishop of Arras, the first body to be dissected at the University of Paris. In almost all cases, the subjects that were dissected were those of executed criminals except for those that were obtained by grave robbing. As early as 1319 four Bolognese medical students were prosecuted for such an act. The Rector of the University eventually issued a directive that no body may be procured for dissection without his permission. Human dissections were regularly carried out at Bologna but it became officially approved by the University in 1405.

AFTER MONDINO

Nicolo Bertuccio (died 1347), who succeeded Mondino at Bologna, continued the practice of human dissection but taught anatomy from his elevated chair. One of Bertuccio's pupils was the famous French surgeon, Guy de Chauliac (1300-70), who continued the Bologna tradition of teaching anatomy at the University of Montpellier in France.

Cooper (1930) reported that dissections were carried out at the University of Montpellier in 1377, 1396 and 1446. From 1340 to 1377 human dissection was permitted once every five years by a special ordinance that had been passed. Prior to this period the teaching of anatomy was limited in scope and even Henri de Mondeville (1260-1320), following his return to Montpellier from Bologna, had to rely on crude anatomical charts, some of which have been incorporated in his text on anatomy published in 1314 (Pagel, 1889; MacKinney, 1962).

The anatomical drawings of de Mondeville, like those of Guido de Vigevano, ushered a new epoch by breaking away from the Alexandrian tradition. In a series of thirteen anatomical miniatures (Ms. fr. 2030, Bibliotheque Nationale, Paris) one sees dissected figures in a standing position demonstrating the skeleton, muscles and various organs. According to Choulant (1920), "de Mondeville's drawings must be regarded as an original accomplishment and his illustrative achievement as very remarkable."

Guido de Vigevano (Gui de Pavie) was physician to King Philippe de Valois of France. He too pursued with great zeal the study of anatomy and compiled, in 1345, an Anathomia, *Liber Notabilium Philippi Septimi*, for the King. The book, which is now preserved at the Musée Condé de Chantilly (Ms. 334/569), contains eighteen large and impressive colored drawings. These figures revealed "for the first time an anatomy in three divisions, the abdomen, the thorax, and the head (Figs. 43-45), an anatomy after the method of Mundinus, whose contemporary and compatriot Guido was and whose pupil he may have been" (Choulant, 1920; Herllinger, 1970).

There are many historical references of human dissection being carried out in other parts of Europe during the 14th century (Lassek,

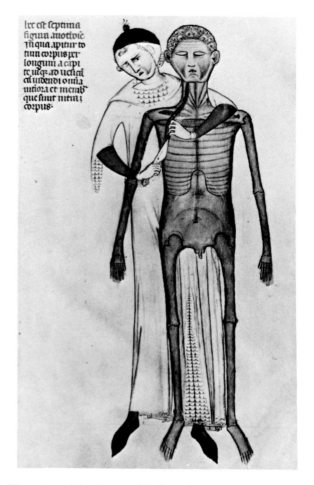

Figure 43. Dissection of the thorax with the cadaver in a standing position. From Guido de Vigevano's Anathomia (Gui de Pavie "Liber Notabilium Philippi Septimi") published in 1345. (Musée Condé de Chantilly Ms. 334/569; Giraudon/ Art Resource, Inc.)

1958; Kevorkian, 1959). These must have been few and infrequent. In almost all cases, the subjects were executed criminals and the dissection was often done with the public participating as spectators. One is left with the impression that the dissection conducted by the anatomists represented the ultimate desecration and punishment of the condemned criminal. In most cases, the law permitted only one

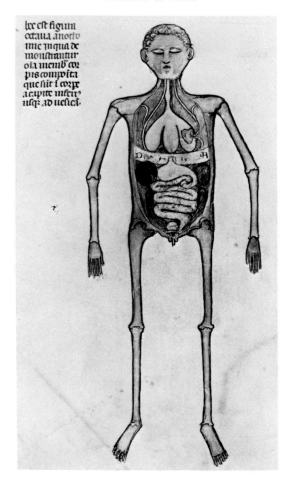

Figure 44. Illustration from Guido de Vigevano's Anathomia (1345) showing a horizontal diaphragm separating the thoracic organs from the abdominal viscera. Almost all the anatomical structures shown are misrepresented. (Musée Condé de Chantilly Ms. 334/569; Giraudon/Art Resource, Inc.)

dissection each year.

With a change in attitude during the latter part of the 14th century, spectacular progress was made in understanding the stucture of the human body.

Figure 45. Rudimentary drawings of two stages in the dissection of the cranium. From Guido de Vigevano's Anathomia (1345). (Musée Condé de Chantilly Ms. 334/569; Giraudon/Art Resource, Inc.)

CHAPTER TEN

LEONARDO DA VINCI

RENAISSANCE ART
AND HUMAN ANATOMY

MANY famous artists of the early renaissance period pursued the study of the human body, including actual dissections, in order to depict the beauty of the human form in an accurate and realistic manner (Töply, 1903, Richer, 1903; Janson, 1977; Parker, 1983). This ferment of art and science brought the study of human anatomy on a new, promising and irreversible course.

Not only was the human body dissected and its skeleton studied, but also more emphasis was placed on the laws of geometry and mechanics as applied to the human form. Indeed, the greatest of these anatomical artists, Leonardo da Vinci, remarked "let no man read me who is not a mathematician. No human investigation can lay claim to being true science unless it can stand the test of mathematical demonstration. The man who undervalues mathematics nourishes himself upon confusion."

Contemporary artists of Leonardo, such as Albrecht Dürer (1471-1528), Antonio Pollaiuolo (1432-98), Verocchio (1435-88), Luca Signorelli (1442-1524), Michelangelo (1475-1564), Titian (1477-1576), and Raphael (1483-1520) followed the same principle but from the standpoint of an artist in order to depict the human form both naturally and accurately by combining artistic ability with keen observation.

LEONARDO DA VINCI

Anatomical Studies

Leonardo also began studying the human body from the standpoint of an artist. It was necessary, not only to visualize the human form, but to understand its deeper structures in order that the surface might be more vividly depicted. Soon Leonardo's enthusiasm for human dissection led him to the study of anatomy for its own sake (Belt, 1955; Keele, 1978, 1983). He studied other problems relating to botany, mathematics, geology, astronomy, and philosophy with the same zeal and in all of these he excelled. He is remembered as the greatest genius of the Renaissance. Leonardo's outstanding accomplishments as an artist, scientist and inventor have been attributed to his originality, creative ingenuity, fertility of invention and his passionate search for new knowledge.

Nowhere does Leonardo's creative genius stand out so clearly as in his magnificent, enthralling and accurate anatomical drawings (Clark, 1935, 1968; Herrlinger, 1953; O'Malley and de C.M. Saunders, 1982). Two years before his death, Leonardo was visited by Cardinal Luis of Aragon and his secretary who recorded the following:

> This gentleman has written of anatomy with such detail showing by illustrations the limbs, muscles, nerves, veins, ligaments, intestines, and whatsoever else there is to discuss in the bodies of men and women, in a way that has never yet been done by anyone else. All this we have seen with our own eyes; and he said that he had dissected more than 30 bodies both of men and women of all ages.

Leonardo was familiar with the authoritative anatomical works of the era. In his notebook he referred to Mondino with reference to the extensors of the toe, in particular the extensor digitorum brevis, which was described by Galen but not mentioned by Mondino. The 1482 edition of Mondino's work was published in Bologna with which Leonardo must have been familiar. Alessandro Benedetti (1450?-1512), a Greek scholar and Professor of Medicine at the University of Bologna and Padua, had published in 1497 an anatomical guide consisting of five books (*Anatomice*). It dealt with the structure and dissection of the human body but lacked originality.

Popular during this period was also the work of the surgeon Guy de Chauliac of Montpellier. His book, *Cyrurgia Magna*, was written in 1363 and a French translation of it appeared in Paris in 1478. Translations of Galen and Avicenna from the Arabic were also available. From his notes, we also know that Leonardo was familiar with Albertus Magnus' (1200?-1280) work in anatomy and zoology, *De Animalibus*.

Leonardo's approach to the study of the human body was that of an astute anatomist (McMurrich, 1930; Keele, 1978, 1983). He recorded in his drawings precisely what he had observed and always attempted to combine structure with function. His dissections were carried out in the hospital of Santa Maria Nuova in Florence and later in Santo Spiritu Hospital in Rome. He dissected in the night, perhaps in secret, and made use of a sharp knife, chisel and a bone saw. There was no method available for preserving the cadaver and he dissected by morselment in order not to damage any of the structures and to expose them as they appeared. His own account of this process is inspiring:

> And you who say that it is better to look at an anatomical demonstration than to see these drawings, you would be right, if it were possible to observe all the details shown in these drawings in a single figure, in, which with all your ability you will not see, nor acquire a knowledge of more than a few veins, while in order to obtain an exact and complete knowledge of these I have dissected more than ten human bodies, destroying all the various members, and removing even the very smallest particles of the flesh which surround these veins, without causing any effusion of blood other than the imperceptible bleeding of the capillary veins. And as one single body did not suffice for so long a time, it was necessary to proceed by stages with so many bodies as would render my knowledge complete; and this I repeated twice over in order to discover the differences. But though possessed of an interest in the subject, you may perhaps be deterred by natural repugnance, or if this does not restrain you then perhaps by the fear of passing the night hours in the company of these corpses quartered and flayed, and horrible to behold, and if this does not deter you then perhaps you may lack the skill in drawing essential for such representation; and even if you possess the skill it may not be combined with a knowledge of perspective, while if it is so combined you may not be versed in the methods of geometrical demonstration, or the methods of estimating the forces and strength of the muscles, or you may perhaps be found wanting in

patience so that you will not be diligent. Concerning which things, whether or not they have all been found in me, the one hundred and twenty books which I have composed will give their verdict yes or no. In these I have not been hindered either by avarice or negligence, but only by want of time. Farewell (Keele, 1952).

From the foregoing passage, it is clear that it was written during the earlier part of Leonardo's life because of the prior reference to having dissected thirty human bodies two years before his death. From the annotations in his notebooks, one learns that he had planned on writing a treatise on anatomy which was probably never completed. The large number of anatomical drawings and extensive notes would have probably found their place in the work he had conceived. He had already mapped out at the beginning of his anatomical studies the scope and content of the book. In his notebook, he had written under the heading "Of the Order of the Book" the following:

> This work should begin with the conception of man and describe the nature of the womb, how the child lives in it, and up to what stage it dwells there, and the manner of its quickening and feeding, and its growth, what interval exists between one stage of growth and another, and what drives it forth from the belly of its mother before the proper time.
> Then describe which are the members which grew more than the others after the child is born, and give the measurements of a child of one year.
> Next describe a grown male and female and their measurements, and the nature of their complexions, color, and physiognomy.
> Afterwards describe how he is composed of vessels, nerves, muscles and bones. This he will do at the end of the book (O'Malley and Saunders, 1982).

Such an expansive plan for a book would have encroached upon Leonardo's time and prevented him from pursuing his diverse interests which included not only anatomy, but the sciences of mathematics, geology, astronomy and philosophy. In 1508 he was already concerned about his voluminous notes, manuscripts and drawings. In his own handwriting, Leonardo recorded "Begun at Florence in the house of Piero Di Braccio Martelli, on the 22nd of March 1508. This makes the collection, without order, taken from many sheets which I have here copied hoping to arrange them later, each in its place according to the matters of which they treat. I believe that before I make an end of this I shall have to repeat the same things many times, for which, O reader, do not blame me, for the subjects are many and the memory cannot retain them and say "This I do not need to write, since I have written it before" (O'Malley and Saunders, 1982).

Two years later the work was still in progress because he himself noted again that he was hoping to finish it in Spring.

For obvious reasons, Leonardo, the artist, must have directed his initial efforts at understanding the musculoskeletal system. His knowledge of mathematics and mechanics is most evident in the numerous drawings showing bones in their correct proportions and limbs in motion. Muscles are revealed in layers and there is no doubt left as to their correct shape and attachments to the bones. Many of the drawings indicate a deep insight as to the opposing functional grouping of muscles and the movement of bones acting as levers, such as in the movement of the arm in pronation and supination and the movements of the bones of the legs and foot during walking (Figs. 46 & 47).

Experiments and Discoveries

Many of Leonardo's discoveries were due to the technical procedures he developed and experiments he carried out both in cadavers and in animals. He demonstrated the shape and extent of the ventricles of the brain in the ox by filling it with melted wax which he poured into it through an opening. He was the first person to use such a technique for demonstrating the size and shape of any body cavity (McMurrich, 1930; Keele, 1979).

The abolition of spinal reflexes in the frog by pitting its spinal column was demonstrated by Leonardo. He carried out experiments on the mechanisms of voice production and described the role of the intercostal muscles during respiration and the events associated with swallowing. His drawings revealed the correct inclination of the pelvis, the papillary muscles, chordae tendineae and the moderator band in the right ventricle.

Leonardo's drawings and writings on the heart, blood vessels and movement of the blood have been extensive and it is quite a surprise that his work did not lead him to discover the circulation of blood. In particular, Leonardo directed his attention to the valves of

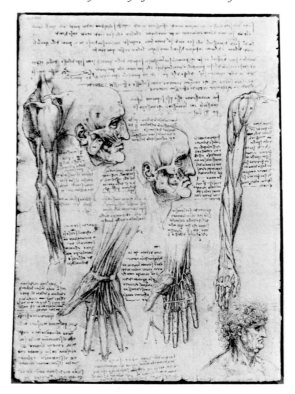

Figure 46. Series of drawings showing the clavicle (and even the subclavius muscle), superficial and deep facial muscles, arm and forearm muscles, the ulnar and median nerves, and the ulnar artery (superficial palmar arterial arch). (Reproduced by Gracious Permission of Her Majesty Queen Elizabeth II.)

the heart and their movement during systole and diastole. His drawings showed that he knew of the origin and distribution of the coronary arteries, as well as of the atria (Fig. 48).

Leonardo knew of the condition of arteriosclerosis (Keele, 1973) as a result of dissecting an old man who had just died. One of his beautiful drawings compared the blood vessels with that of a younger person.

> The artery and the vein which in the old extend between the spleen and the liver, acquires so great a thickness of skin that it contracts the passage of the blood that comes from the mesenteric veins, through which this blood passes over to the liver and the heart, and to the two greater vessels, and in consequence to the

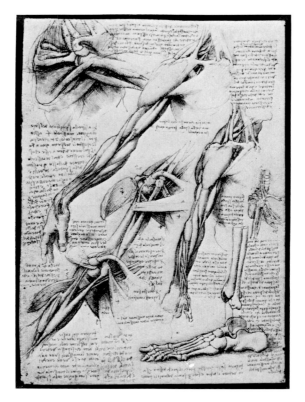

Figure 47. Leonardo's drawings of the muscles of the shoulder region, arm and forearm. The bones of the ankle region and foot are shown below on the right. (Reproduced by Gracious Permission of Her Majesty Queen Elizabeth II.)

whole body. And apart from the thickening of the skin (or coat) these veins grow in length and become tortuous like a snake, and the liver loses the humor of the blood which was carried there by this vein, and consequently this liver becomes dried up and becomes like a congealed bran (crusca congelata) both in color and substance (Keele, 1952).

Leonardo was the first to depict the appendix in his implicit drawings of the gastrointestinal tract (Fig. 49). However, he conceived the diaphragm and muscles of the abdominal wall as the forces controlling movement of the gut. His lack of appreciation of peristalsis in the wall of the gastrointestinal tract was evident too in his description of the ureters and of the flow of urine from the kid-

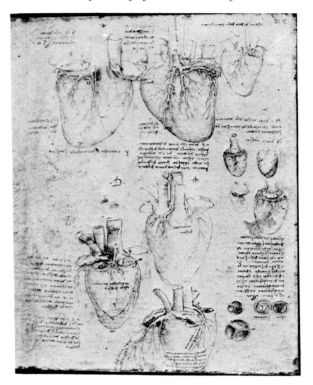

Figure 48. Sketches showing the distribution of the coronary vessels, viewed from different aspects. On the lower right hand corner are the pulmonary and tricuspid valves seen from above. Leonardo's observations were based on dissection of the ox heart. (Reproduced by Gracious Permission of Her Majesty Queen Elizabeth II.)

neys to the bladder. He viewed the ureter as a simple tube through which fluids flowed as a result of gravity and even demonstrated in a series of diagrams the effect of various bodily positions on the flow of urine from the kidneys to the bladder.

For an understanding of early development he studied incubated chick eggs and the embryos of lower mammalian animals. The placenta and fetal membranes were drawn and described. Unlike his predecessors, Leonardo drew the uterus as containing a single cavity, complete with the ovaries and the uterine tubes, as well as their related blood vessels. One of his magnificent drawings shows a fully developed fetus within the bisected uterus with the umbilical cord

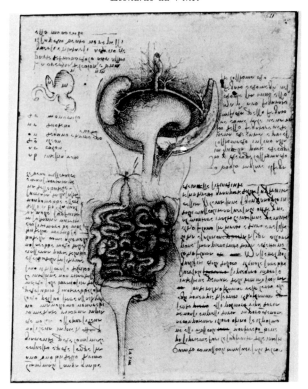

Figure 49. Drawing of the gastrointestinal tract, showing for the first time the vermiform appendix. Above, the esophagus, stomach and intestine are depicted; below lie a shrunken liver, enlarged spleen and a prominent splenic vein in relation to the stomach. The vermiform appendix is represented in the diagram above and again more prominently on the lower right. (Reproduced by Gracious Permission of Her Majesty Queen Elizabeth II.)

attached to the placenta (Fig. 50).

Anatomical Masterpieces

In more than 750 anatomical drawings of the muscular, skeletal, vascular, nervous, and urogenital systems, Leonardo produced a memorable body of work of unchallenged artistic beauty and scientific accuracy. He depicted parts of the human body from different perspectives and to him we owe the first cross-sectional drawings in anatomy. For Leonardo it was not enough to illustrate but also to

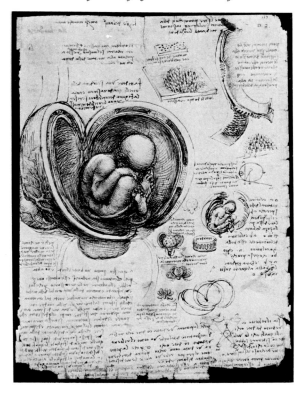

Figure 50. The pregnant uterus bisected to show the fetus in its natural position. The placenta, however, is that of a cow. The sketches on the right and below reveal the different layers of fetal membranes. In one of the small diagrams, the fetus can be seen through the transparent amnion following removal of the chorion and uterine wall. (Reproduced by Gracious Permission of Her Majesty Queen Elizabeth II.)

explain and his drawings were often accompanied by questions and remarks relating to functions (Fig. 51 & 52). So much that he had discovered and recorded remained unknown for centuries to emerge again only in relatively recent years. Much of what he had accomplished was not to be surpassed for centuries.

Leonardo's anatomical drawings were in the tradition that began with Giotto which displaced conventionalism and aimed at a more natural and realistic representation. But to create such magnificent anatomical drawings, demanded not only the skill to sketch accurately, but the unique ability of meticulous dissection and represen-

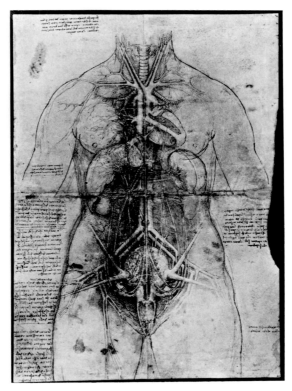

Figure 51. This elaborate and impressive drawing is undoubtedly an earlier work of Leonardo. There are many inaccuracies and the anatomical representations suggest that the drawing was based on the dissection of animals. An attempt has been made to consolidate the structure of the female anatomy with Galenic concepts of function. (Reproduced by Gracious Permission of Her Majesty Queen Elizabeth II.)

tation of the structures that were displayed. Not satisfied with descriptive anatomy he forged ahead to explain the function of various structures.

Leonardo died on the 2nd of May, 1519. He bequeathed all his manuscripts and drawings to his beloved disciple and friend, Francesco Di Melzi, who kept them secured for almost half a century in admiration of his great master. Following the death of Melzi in 1570, the manuscripts were passed on to his nephew Orazio who could not have appreciated their value. The manuscripts eventually were dismembered and dispersed, passing through many hands, including

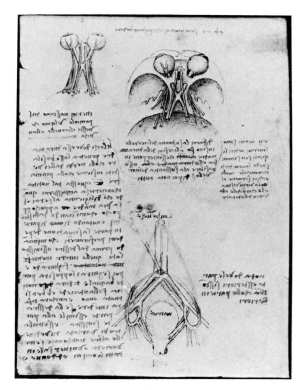

Figure 52. The sketches above show the optic chiasma and several cranial nerves. Below is a drawing of the uterus with paired vessels supplying it from either side. (Reproduced by Gracious Permission of Her Majesty Queen Elizabeth II.)

Pompeo Leoni and Thomas Howard, Earl of Arundel. Several facsimile editions of Leonardo's anatomical drawings have been published but most of the originals are now in the Royal Library at Windsor Castle (Clarke, 1968; Keele, 1979; O'Malley and Saunders, 1982).

Despite errors and misconceptions, Leonardo did make significant discoveries in anatomy. Not only did he describe certain anatomical structures for the first time, but he also recognized the true curvature of the spinal column (Fig. 53) and the true position of the fetus *in utero*. It remains, however, undisputed that Leonardo's work is of epochal importance. Unlike his contemporaries, it was Leonardo alone who pursued the study of the human body with such

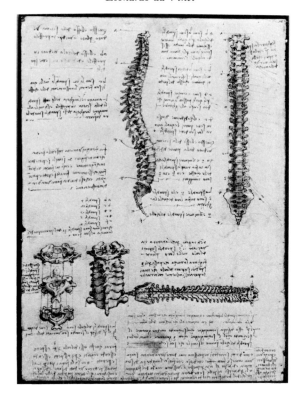

Figure 53. Leonardo's drawings of the articulated vertebral column (viewed from the lateral, anterior and posterior aspects). The spinal curvatures are evident. (Reproduced by Gracious Permission of Her Majesty Queen Elizabeth II.)

thoroughness that he quickly transcended the needs of the artist and drifted into the scientific pursuit of anatomy for its own end. His disciplined mind and scientific rectitude brought human anatomy into its first modernization phase. That his extraordinary and dazzling anatomical masterpieces remained hidden and unpublished for more than three centuries can only be described as a surpassing irony in the life of this accomplished and inspired genius.

CHAPTER ELEVEN

MEDIAEVAL ANATOMISTS

SCHOLASTICISM

STULTIFIED in the spirit of scholasticism the study of human anatomy as a practical and intellectual discipline languished for another two centuries. It was, however, not an entirely barren period because human dissections were often carried out in many Italian cities and elsewhere in Europe, even though lacking an enquiring and critical approach. The successors of Mondino commented on his treatise, the *Anathomia*, which recognized anatomy as a distinct discipline and not part of surgery. Lind (1975) revealed through his exhaustive research of this period the true contributions and influence of the work of these scholars on the evolution of anatomical knowledge.

According to Lind (1975) mediaeval scholasticism and Renaissance humanism provided the cultural background for anatomy to advance during this period. Under such influences, it became possible "at the two great centers of mediaeval and Renaissance Italian university life, Bologna and Padua, where gradually more facilities for scientific investigation could be found, where both papal and Venetian liberalism had long ago removed restriction upon dissection, and where the great medical minds of the time both in Italy and behind the Alps gathered to establish a tradition of free enquiry and research." This undoubtedly led to advances in other related branches of medicine. Thus pathological anatomy emerged with the publication in 1507 of Antonio Benivieni's (Fig. 54) book, *De Abditis*

Nonnullis ac Mirandis et Sanationum Causis. This work described for the first time post-mortem findings from autopsies carried out in an attempt to explain causes of symptoms and to determine the organs responsible for the diseases.

Lind (1975) divided the mediaeval anatomists into two groups, the forerunners and their successors. The first group included Hieronymus de Manfredi, Alessandro Achillini, Alessandro Benedetti, Gabriele Zerbi, and Berengario da Carpi.

The second group included Antonio Benivieni, Andres de Laguna, Niccolo Massa, Johann Guenther of Andernach, Johannes Dryander, Giovanni Battista Canano and Charles Estienne. Of im-

Figure 54. Antonius Benivieni (1443-1502). A contemporary of Leonardo da Vinci, he published the first book on pathological anatomy. (With permission, from Major, 1954.)

portance too were the following: Jacobus Sylvius, Johann Peyligk and Magnus Hundt. Undoubtedly, the most influential of these was Berengario da Carpi whose personal and extensive contributions have helped to advance anatomy in the proper direction, so much so that he is considered as "the first truly original mind in anatomical research since Mondino" (Lind, 1975).

Giacomo Berengario (Fig. 55) was born in Carpi in 1470. He studied medicine at the University of Bologna, where he taught anatomy and surgery between 1502 and 1527. After a lucrative practice in Rome and a lengthy retirement in Ferrara, he died in 1550.

Berengario wrote two anatomical works, the extensive *Commentaria Super Anatòmia Mundini* in 1521 in Bologna (Fig. 56), and the much smaller *Isagogae Breves* which appeared a year later. In an attempt to improve Mondino's book, Berengario made corrections and added illustrations (Figs. 57 and 58) which were based on actual

Figure 55. Berengario da Carpi (1470-1550). (With permission, from Major, 1954.)

human dissections. Apparently, he had carried out more than one hundred human dissections and was even accused of vivisection. The major book contained twenty-one wood-cut plates, even though crude, were far superior to the earlier anatomical drawings (Figs. 59-61) which were in circulation at that time. He wanted his illustrations to be of use not only to doctors but also to artists.

The first account of the basilar part of the occipital bone, the sphenoidal sinus and the tympanic membrane was provided by Berengario. He gave an excellent description of the abdominal muscles, greater omentum, transverse colon, vermiform appendix, renal papillae, tricuspid valve, ventricle of the heart and semilunar valves of the pulmonary artery. He also mentioned the pharyngeal cartilages and cochlea. According to Galen openings were present in the septum of the heart, but Berengario tactfully stated that these were

Figure 56. Facsimile of the title page of Berengario's elaborate commentary (1521) on the work of Mondino. (Courtesy of the Royal College of Physicians, London.)

Figure 57. Musculature of the human body. This full-length plate of the dissected muscles viewed from the back is less schematic than earlier illustrations. (From Berengario's *Isagogae Brevis*, Bologna, Benedictus Hectoris, 1523; from Major 1954, with permission.)

seen with great difficulty in the human.

GABRIELE DE ZERBI

Gabriele de Zerbi (1445-1505) taught medicine and philosophy at the Universities of Padua and Bologna. His work, *Anatomiae Corporis Humani et Singulorum Illius Membrorum Liber*, was published in Venice in 1502. It is essentially a commentary on Mondino's treatise which was the pre-eminent anatomical work of the era.

According to Lind (1975) Zerbi's book was "the first systematic and sufficiently detailed examination of the human body since Mundinus and far outstripped the latter in scientific accuracy" with

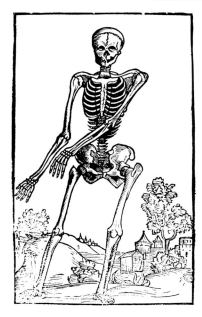 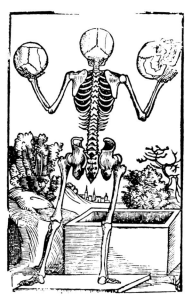

Figure 58. One of the first serious attempts to depict the human skeleton in a printed book. From Berengario's *Isagogae Brevis*, Bologna, 1522. The same plates had been used by Berengario in his edition of Mondino's Anatomy in 1521. (From Cushing, 1943; with kind permission of the Blackie Publishing Group, Glasgow.)

numerous quotations from the established medical literature of the Greek and Arab scholars. The format, language, style and abbreviations followed that of Mondino. For each organ or part of the body, the following nine criteria were adopted: substantia, numerus, complexio, figura, quantitas, situs, colligantia, iuvamentum, passiones. Occasionally, color and operations were added. In addition to his extensive commentary on Mondino's work and quotations from the work of other scholars, Zerbi also recorded in his book personal observations with respect to the uterine tubes and the lacrimal gland even though he described two for each orbit.

ALESSANDRO ACHILLINI

Alessandro Achillini (1463-?1512) was both a philosopher and

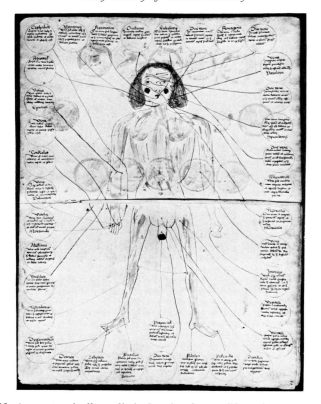

Figure 59. An anatomically realistic drawing from a fifteenth century manuscript showing blood-letting points. The nude body is boldly depicted with muscles, external genitalia and other distinct features (Vatican, Ms. Palat. Lat. 1709, fol. 44v-45r; Foto Biblioteca Apostolica Vaticana). Bloodletting was carried out in order "to reduce or to correct excessive or corrupt humours" and based on Greek and Roman healing practices. It complemented cauterization and specific places on the body were identified (as shown on the figure) for various ailments (Mac-Kinney 1965).

physician in Bologna (Fig. 62). For a full description of his life and anatomical studies, reference should be made to the works of Münster (1933), Matsen (1969), and Lind (1975). According to Lind (1975), Achillini taught medicine and philosophy at the University of Bologna for twenty-six years, longer than anyone else. His eulogy indicated that "he was so skilled in all sciences that he was by the unanimous opinion of all the learned placed beyond all hazard of genius. There was in him a singular vigilance, energy, and activity." He

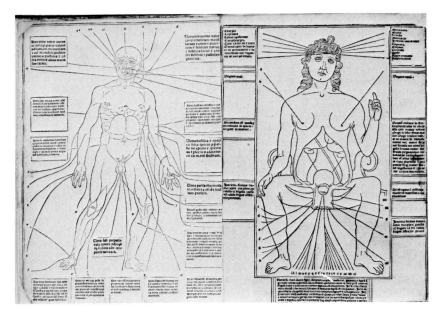

Figure 60. Anatomical woodcuts from the 1500 Venice edition of Mondino's *Anathomia*. The figure of the man on the left shows the locations for bloodletting. The female reproductive organs are depicted on the right. (By permission of the British Library.)

was held in the highest esteem by his colleagues but at the same time envied by many on account of his ability and success.

Achillini's major work on anatomy *Anatomice, Siue Historia Corporis Humani* was published in 1502 and contained many of his anatomical discoveries which include the duct of the sublingual gland, the vermiform appendix, the cecum, the iliocecal valve, and at least two of the auditory ossicles (malleus and incus). His best known book *Annotationes Anatomicae* was published in 1520 (Fig. 63) after his death. Various parts of the body have been described following the format of Mondino but without any extensive commentary on his work. Achillini was extremely familiar with the brain and the bones of the body and in his book he compared his own observations with those of other authorities.

De origine rerū naturaliū

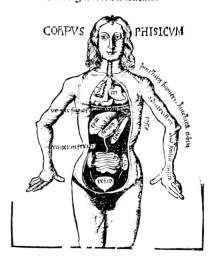

Figure 61. Woodcut of the dissected thoracic and abdominal cavities, from Gregor Reisch (c. 1467-1525) encyclopedical *Margarita philosophica*, published in 1503. (With permission from Major, 1954.) The diagram reveals the following: the trachea in the neck; in the thorax, the right lung and on the left side, the heart; diaphragm; liver, stomach, spleen, intestine, kidneys and bladder in the abdominal cavity. *Corpus phisicum* is inscribed above the figure. In subsequent editions the *situs* figure was copied and modified.

NICCOLO MASSA

The Venetian Niccolò Massa (1485-1569) is remembered for his book, *Liber Introductorius Anatomiae Siue Dissectionis Corporis Humani, etc.*, which was first published in 1536 and reissued in 1559 (Fig. 64). Lind (1975), who translated this work, considered it to be "one of the more modern 16th century anatomies" because Massa "abandoned the old narrow practice of assembling authorities and choosing alternatives among conflicting opinions." Furthermore he displays common sense and a vigorous mind and thrusts aside all shame and ignorance: his attitude towards slavish adherence to authority is expressed in the following statements:

> Let it not be ascribed to my ignorance if I point out parts of the body not seen by Aristotle and Galen and many others, since it is given to each person to seek better things to the extent of his ability. For questions and difficulties not solved by Averroes as well

Figure 62. Alessandro Achillini (1463-?1512). (Wellcome Institute Library, London.)

as other wise men have made me a sedulous investigator of the parts of the human body and of other things.

Like many of his contemporaries, Massa's book followed the arrangement of Mondino. He recognized the need for human dissection which he often carried out, including on the bodies of stillborns. It is a verbose work but at the same time presented a clear and systematic description of the "many parts, functions, and uses as possible of the human body which had been thus far overlooked by others both ancient and more recent and now for the first time made manifest; a work quite useful indeed to all students of medicine and philosophy" as is indicated on the title page of the first edition (Lind, 1975).

Massa incorporated a fair amount of his own observations from human dissections into the prevailing Galenic teachings. There were really no new discoveries mentioned and it seems that the work had no great influence at that time except for the celebrated French physician and anatomist, Jacobus Sylvius (1478-1555), who made frequent references to it in his book the *Isagogae*.

Hannibal Camillus Corrigiensis artium & Medicinæ diſcipulus
Lectori.   S.   D.
Congeriem ſtudioſa cohors fabricantia noſce
Corporis humani femina ſi qua uelit.
Verſet Achilleum noctelq; dieſq; uolumen.
Quod nimis exiguum ne tamen eſſe putet.
Quicquid enim eximii toto deus orbe creauit
Paruulus in parua molle recondit homo.

Eiuſdem Hanniballis Diſticum.
Omniparens hominem fabrefecit uernula ſummi
Patris. Alexander conſcius illa docet.

Figure 63. Title-page of Alessandro Achillini's book, *Anatomicae annotationes*, Bologna, de Benedictis, 1520. (With permission, from Major 1954.)

ANDRES DE LAGUNA

Andres de Laguna (1499-1560), born in Spain, graduated from the University of Toledo in medicine. He spent several years in France including one year (1532-1533) in the medical faculty of the University of Paris. He published a small work entitled *Anatomica Methodos, Seu De Sectione Humani Corporis Contemplatio* in 1535. It was essentially an unremarkable work which revealed the influence of Galen, Hippocrates and the classical Greek scholars such as Plato, Pliny, and Aristotle.

With respect to his views on personal observations and dissections, as well as on the work of Mondino, this is what Laguna had to say (Lind, 1975):

He who wishes to recognize accurately the care and the industry of nature must present himself as a dissector of even the repulsive bodily members and most diligently examine their location, shape, number and consistency. Once at Paris I attended an

Figure 64. Niccolò Massa (1485-1569). (Wellcome Institute Library, London.)

anatomy presided over by all the fellow practitioners of the medical art as well as the barbers to whom the actual task of dissection was committed. In order to avert from themselves the stink of the intestines and to leave the matter as one known to require explanation they declared without even looking at them that the cecus intestine had only one orifice. But I took out a scalpel, dissected this intestine and demonstrated to all with a small peg or stick that it had two small openings in line with each other, one through which it attracted and another through which it expelled the feces. For I had read in Mundinus, which is not so ignorant as he is barbarous, that such was the fact, as I discovered with my own eyes.

The single illustration in the book was copied from that of Pietro d'Abano's *Conciliator* which appeared in 1496. Laguna admonished his students that if they are not skilled in anatomy they should study this diagram in order to learn about the structure and arrangement of the muscles of the anterior abdominal wall. The illustration was crude, fanciful and useless.

Laguna stated that the abdominal wall has no bone "since some times it must subside as when it is without food; very often, however, it swells with a great amount of food or with the conception of a fetus, as in pregnant women. In these cases bone would be a great hindrance. . . . Likewise in the coitus of male and female a mutual collision would take palce if the stomach of each sex were bony; hence everyone would abstain from the act of generation as much as possible and the entire human race would perish very shortly" (Lind, 1975).

GIOVANNI CANANO

The muscles of the extremities were studied in much detail from actual human dissections by Giovanni Battista Canano (1515-1579). Canano (Fig. 65) was born in Ferrara where he studied medicine and later became Professor of Anatomy. He assumed this position in 1541 having succeeded his relative Antonio Maria Canano who was a student of Marcantonio dalla Torre at the University of Padua. Canano's work *Musculorum Humani Corporis Picturata Dissectio* was a small volume of 20 leaves with 27 impressive illustrations based on human dissections. The date of publication of this volume is not known and apparently only few copies were then in circulation. According to Choulant (1852, 1920) the illustrations were unusually accurate and the book was probably printed before 1543. The work is incomplete and of the others mentioned in the preface, as "presently going to publish the remaining which are already in press," nothing is known.

John Caius visited Canano in 1543 and was most impressed by his collection of books. Canano is credited with the discovery of valves in the veins which were later to be rediscovered by Fabricius in 1574. He demonstrated the palmaris brevis muscle to Gabriele Fallopio who at the age of twenty-four was appointed Professor of Anatomy at Ferrara.

CHARLES ESTIENNE

The French anatomist, Charles Estienne (?-1564) is credited with many anatomical discoveries including the synovial glands, the spinal canal, and the capsule of the liver. His book *De Dissectione Partium Corporis Humani* appeared in Paris in 1545. The work seemed to

Figure 65. Giovanni Battista Canano (1515-1579). (Wellcome Institute Library, London.)

have occupied him over a relatively long period and he was assisted by a surgeon, Etienne Riviere, who aparently did many of the dissections and drawings from the specimens.

Descended from a distinguished family of printers it is quite natural that Estienne would have paid particular attention to the quality of the illustrations. The book consisted of sixty-two full page plates, including several repetitions, and many engravings in the text itself. Because of the excellent wood cut engravings, the illustrations conveyed artistic merit but at the same time lacked anatomical accuracy. The anatomical positions were unusual and parts of the body were poorly presented. In contrast, the text accompanying the illustrations was clear and instructive because of the many personal observations he made during dissecting.

JOHANNES DRYANDER

Johannes Dryander (Fig. 66) whose original name was Eichmann, was born about 1500 in Oberhessen, Germany. He studied mathematics and astronomy in Erfurt and later medicine in Paris and Mainz where he received his Doctor of Medicine degree in 1533. Returning to Germany, he became Professor of Medicine and Mathematics at the University of Marburg.

Dryander encouraged human dissection and taught anatomy as a practical discipline from 1536 to the time of his death in 1560. His public dissections were among the first in Germany and he was also one of the first anatomists to make illustrations from his own dissections.

Dryander was greatly influenced by both Mondino and Berengario and the five years he spent in Paris (1528-1533) coincided with the time both de Laguna and Vesalius were students there. His

Figure 66. Johannes Dryander (c.1500-1560). (Wellcome Institute Library, London.)

extensive publications included two works in anatomy which were illustrated with drawings from his own dissections. At the same time
he made generous use of illustrations from the work of Berengario
and Vesalius.

Of the forty-five figures and drawings (Figs. 67-69) in Dryander's *Anatomia Mundini*, which was published in Marburg in 1541,
eighteen plates were modified copies from Berengario's. The work
itself consisted of sixty-seven pages and most of the plates occupied
the full page. His *Anatomiae, Hocest, Corporis Humani Dissectionis Pars*
. . . was also published in Marburg in 1536, a year after his previous
book *Anatomia Capitis humani*. Some of the illustrations were taken
from the work of Phryesen's *Spiegel der Artzney* and from Vesalius'
Tabulae Sex, but many of his own drawings were also included.

Dryander's illustrated anatomical works marked an important
milestone in the history of anatomical illustrations. It should be
borne in mind that although human dissections were carried out in
the 14th century and anatomical works were being published, there
seems to have been no great need for anatomical illustrations as evi-

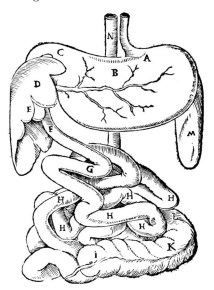

Figure 67. Diagram of abdominal viscera. From Johannes Dryander (Johann
Eichmann) Anatomia Mundini, 1541 (By permission of the British Library). The
esophagus, stomach, small and large intestine, the liver and spleen are shown. Although the organs appear life-like, there are many inaccuracies.

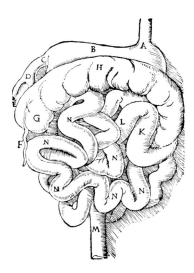

Figure 68. Drawing of the gastrointestinal tract. It is considered to be the earliest printed figure showing the vermiform appendix. (From Johannes Dryander's *Anatomia Mundini*, Marburg, Egenolph, 1541; with permission from Major 1954.)

denced even in the famous work of Mondino. The situation remained unchanged even with the printing of books and the widespread use of woodcuts during the second half of the 15th century. The schematic and often fanciful representations of human anatomy as seen in the works of Johannes Ketham, Johannes Peyligk, Magnus Hundt and many others (Figs. 70-77) were of little value to physicians (Choulant, 1852; Locy, 1911; Garrison, 1915; Crummer, 1923; Lint, 1924).

In circulation at that time was a series of five schematic pictures (Fünfbilderserie) illustrating the skeletal, nervous, muscular, venous and arterial systems in squatting positions. Often, a sixth drawing showed the reproductive system or a pregnant woman. These drawings were crude and imaginative. As revealed by recent studies these drawings had been transmitted through the centuries and might have been influenced by physicians during the period of antiquity (Sudhoff, 1907, 1908, 1910; Garrison, 1915; Wegner, 1939; O'Neill, 1977).

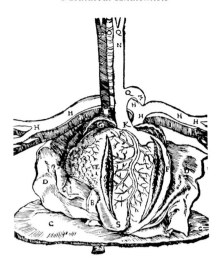

Figure 69. Illustration of the heart, pericardium and great vessels. (From Johannes Dryander's *Anatomia Mundini*, Marburg, Egenolph, 1541; with permission from Major 1954.)

JOHANN PEYLIGK

Johann Peyligk (1474-1512), Professor of Law, was the author of *Compendiosa capitis physici declaratio* . . . . and of a highly philosophical work, *Philosophie Naturalis*, published in 1499 in Leipzig. The last chapter of this book is devoted to human anatomy and contains several woodcuts showing the body cavities, viscera and various organs (Figs. 78-81). The drawings are highly schematic and crude (Sudhoff, 1916; DeFanu, 1962).

MAGNUS HUNDT

The anatomical woodcuts in Magnus Hundt's book, *Antropologium de hominis dignitate, natura et proprietatibus, de elementis, partibus, et membris*, was printed in Leipzig in 1501. The work consists of 120 pages and is illustrated by several woodcuts of the head, face, neck, cranial sutures, brain, cerebral ventricle, palm of the hand, and a complete skeleton, as well as one showing the thoracic and abdominal organs (Figs. 82-84). Although not the oldest of anatomical

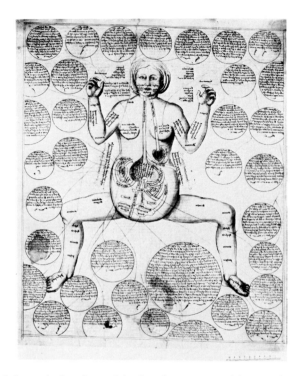

Figure 70. Schematic drawings of the female anatomy with extensive annotations of diseases in the captions surrounding it (c.1515-1520). (Ms. Ny. Kgl. Saml. 84b. 2°, f. 2r; Det Kongelige Bibliotek, Copenhagen.) The extremities, face and various thoracic and abdominal organs have been labelled. The typical Alexandrian squatting posture with raised arms, bent at the elbow, is also seen here. Choulant (1920) regarded these drawings as anatomically worthless, but at the same time providing an important link in the evolution of illustrations depicting the *gravida*.

illustrations, they are for that period the most complete representation of the internal organs as envisioned during the 15th century.

LAURENTIUS PHRYESEN

Much improvement is seen in the treatise of the Dutch physician, Laurentius Phryesen. His *Spiegel der Artzny* was first published in Strassburg in 1518. The anatomical illustrations it contains are

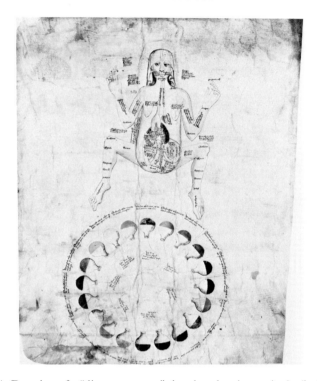

Figure 71. Drawing of a "disease woman" showing the viscera *in situ* (before 1450) (Ms. Ny. Kgl. Saml. 84 b, 2°, f. 4r; Det Kongelige Bibliotek, Copenhagen). Mediaeval artistic representations of the human figure in the traditional squatting posture. The internal organs appear to be drawn on the surface of the body. The uterus is on the left with a fetus standing in it. Below is a chart giving instructions for the diagnosis of diseases by uroscopy.

dated 1517 and have been attributed to Johannes Waechtelin, a pupil of the elder Holbein.

The first plate is the work of Wendelin Hock, who practiced medicine in Strassburg, and was published by Johann Schott of Strassburg in Gerssdorff *Feldtbuch der Wundtartzney* in 1517. It shows the open thoracic and abdominal cavity separated by a horizontally placed diaphragm (Fig. 85). There are two lobes in the right lung and the liver is shown to consist of five. Both the stomach and bladder are depicted as spherical structures and the kidneys are placed high up and connected to the bladder by very short ureters. Intriguing are the diagrams around the body showing different stages in

Figure 72. Woodcut of common wounds on an imaginary representation of the abdominal viscera (From Hieronymus Brunschwig's *Buch der Cirurgia*, published in 1497, Strasbourg; Courtesy of the Medical Library, University of Manitoba). An illustration of a similar "wound-man" also appeared in several other surgical works including Ketham's Fasciculus medicinae, Venice 1493).

the dissection of the brain, as well as the tongue. The structures are identified by German names. Representation of the anatomical structures seen in Fig. 73 are essentially similar to those in the previous illustration.

The plate of the articulated skeleton (Fig. 86) was also first published by Johannes Schott in 1517. These illustrations with all their shortcomings are unique with respect to presentation and originality.

About 300 B.C. in Alexandria anatomy was first taught as a practical subject through actual dissection of the human body. Because of Galenism this tradition was lost until the early 14th century. Even Mondino and his contemporaries were content to demonstrate and verify Galenic teaching when dissecting the human body. Anatomy was revolutionized and medicine once again became a modern

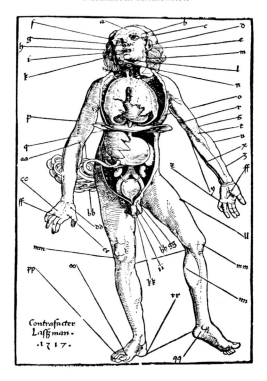

Figure 73. Thoracic and abdominal viscera. (From Gerssdorf's *Feldtbuch der Wundtartzney*, 1517. Medical Library, University of Manitoba.)

scientific discipline, both at the same time, with the publication in June 1543 of Vesalius' *De Humani Corporis Fabrica* (Feyfer, 1914; Sigerist, 1922; Dempster, 1934; Underwood, 1943). We will now turn our attention to the genius that has produced it.

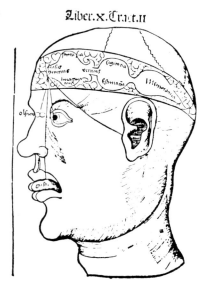

Figure 74. Drawing of the three "cerebral cells" from the *Margarita philosophica*, 1503, of Gregor Reisch (c. 1467-1525) (with permission, from Major, 1954). The organs of special senses are connected to the anterior cell, labelled *Sensus communis, Fantasia* and *Imaginativa*. The narrow canal between the anterior and middle cells is designated vermis. The middle and posterior cells were labelled *Cogitativa* and *Estimativa*, respectively. The localization of the senses in the cerebral ventricles was a mediaeval concept.

Figures 75-77. These illustrations of the *foetus in utero* are from a twelfth-century manuscript, derived from the work of Soranus of Ephesus, a second century A.D. Roman physician. The fetuses are shown in various positions, surrounded by chorion, within a flask-shaped uterus. The double circles represent the peritoneum. The drawings include one of a twin pregnancy (Ms. Gl. Kgl. Saml. 1653, 4°, f. 17r-19v; det Kongelige Bibliotek, Copenhagen). The illustrations in Eucharius Roslin's *Rosegarten* (1513) and William Raynalde's *Byrthe of Mankynde* (1545) were adapted from Soranus' drawings and from the ninth century Moschion codex (3701-3714) in the Royal Library, Brussels. Leonardo da Vinci's drawing of the Foetus *in utero* is the first realistic representation of the natural position of the fetus.

Figure 76.

Figure 77.

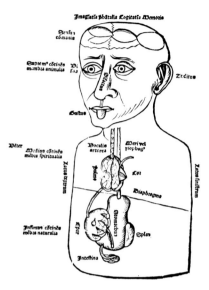

Figure 78. Diagram from Johannes Peyligk's *Philosophie Naturalis Compendium*, published 1499 in Leipzig (By permission of the British Library). The ventricles of the brain and the organs in the thorax and abdomen are crudely represented. The diaphragm is indicated by an oblique line. The heart is conical in shape with a large vessel leading to the liver and projects into the right side of the thoracic cavity; the liver is five-lobed and the lower end of the intestine is peculiarly knotted.

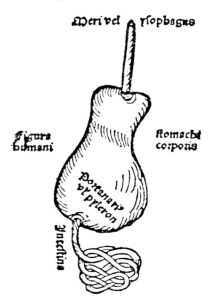

Figure 79. Drawings of the cerebral ventricles, within the skull (above) and viewed laterally (below). The two lateral ventricles are depicted adjacent to each other for the first time. The ventricles are shown in communication with the hypophysis (lacuna cerebra). (From Peyligk's *Philosophie Naturalis Compendium*, Leipzig, Melchior Lotter, 1499.)

Figure 80. The gastrointestinal tract as depicted in Peyligk's *Philosophie Naturalis Compendium* (Leipzig, Melchior Lotter, 1499).

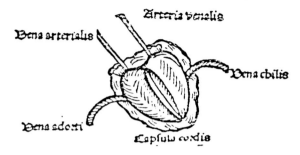

Figure 81. The heart and major blood vessels (From Peyligk's *Philosophie Naturalis Compendium*, Leipzig, Melchior Lotter, 1499).

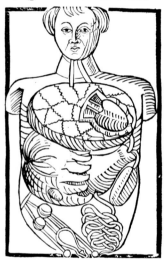

Figure 82. Diagram of the thoracic and abdominal organs. From Magnus Hundt's *Antropologium*, Leipzig 1501. (By permission of the British Library.) Observe the five-lobed liver, paired kidneys on the right side of the abdomen, spleen on the left side with a duct leading into the stomach, and intestines that appear to be knotted. The heart and lungs are schematically depicted.

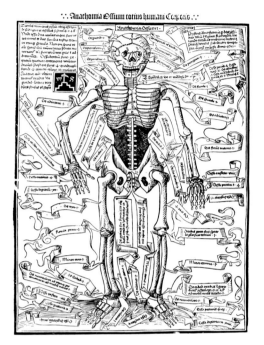

Figure 83. Diagram of the human skeleton in Magnus Hundt's *Antropologium*, 1501. (By permission of the British Library.) This illustration is probably derived from the print attributed to Richard Helain (Nurenberg 1493). Observe the opened pelvis, poor representation of the lumbar vertebrae, and the peculiar sutures of the skull. Nevertheless, it is a great improvement when compared with earlier drawings of the skeleton, such as in the Provencal manuscript (13th century, Ms D.11, 11, folio 169v, Universitätsbibliothek, Basel), Dresden manuscript of 1323 (see Sudhoff, 1908), and the 14th century codex in Munich (Staatsbibliothek, Cod.lat. 13042).

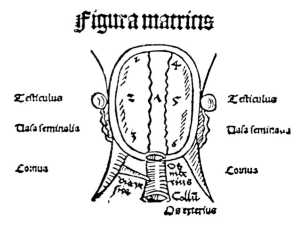

Figure 84. Drawing of the uterus, shown divided into seven compartments. (From Magnus Hundt's *Antropologium*, Leipzig 1501; courtesy of the Royal College of Physicians, London.)

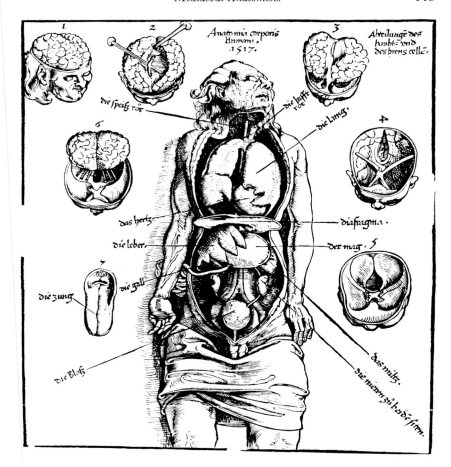

Figure 85. Illustration of the thoracic and abdominal viscera, from Gerssdorf's *Feldtbuch der Wundtartzney*, 1517. Various stages in the dissection of the brain are depicted around the figure. (Medical Library, University of Manitoba.)

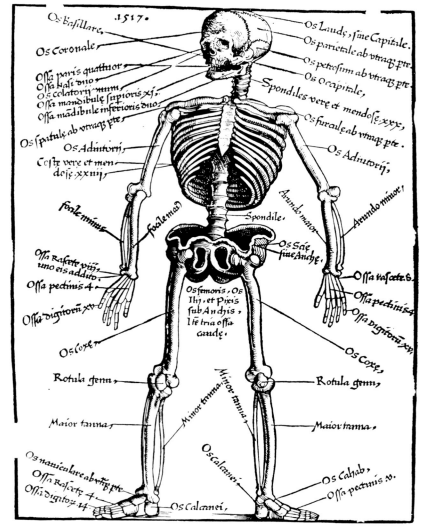

Figure 86. Diagrammatic representation of the human skeleton. (From Gerssdorf's *Feldtbuch der Wundtartzney*, 1517. Medical Library, University of Manitoba.)

CHAPTER TWELVE

ANDREAS VESALIUS —
ARCHITECT OF THE NEW ANATOMY

FAMILY AND EARLY EDUCATION

ANDREAS Vesalius, "the most commanding figure in European medicine between Galen and Harvey" (Garrison, 1929) was born on December 31, 1514 in Brussels, Belgium. He came from a distinguished family of physicians, whose name was originally Witing and who later adopted the name Wesalius after the town of Wesel on the Rhine from where they had originated. His father, Andreas Witing, was pharmacist to Margaret of Austria and later to her nephew, Emperor Charles V. The great-great-grandfather of Vesalius, Peter, was a physician who wrote a treatise on the fourth Fen of Avicenna. He left a large and valuable collection of medical works, some of which were preserved by Vesalius' mother and were later to be treasured by him during his student days.

John, the son of Peter, became Professor of Medicine at the University of Louvain and, according to the records of the University, served on many of its administrative committees. John was appointed Physician to the City of Brussels and later entered the services of the Duke of Burgundy. From Frederick III he received the heraldic emblem depicting the three Weasels which was also that of the town of Wesel. It became the family's coat of arms and was later to adorn the title page of Vesalius' magnificent treatise.

Everard, John's eldest son, studied medicine at the University of Louvain. He was physician to Mary of Burgundy and later to the

Emperor Maximilian I after her marriage to him. Everard wrote commentaries on the books of Rhazes and on the first four sections of the Aphorisms of Hippocrates. It has been said that the former work might have inspired Vesalius in the writing of his *Paraphrase of the Ninth Book of Rhazes* which was published in 1537 at Louvain as a requirement for his Baccalaureate thesis. Vesalius was therefore to pursue a career, steeped in his family tradition, which was imbued with scholarship, medical matters and service to the aristocracy.

Between 1528 and 1533, Vesalius studied at Louvain at the Pedagogium Castri and later at the Collegium Trilingue. He learned Greek, Latin and Hebrew in the best humanistic tradition because at that time knowledge of the ancient languages was an essential pre-requisite for all scholarly pursuits.

We learn from Vesalius himself that during his early years he became interested in the mediaeval works of scholarship and in anatomy. It was during this period also that he began to study anatomy in a very practical way by dissecting mice, moles, rats, dogs and cats. He referred to one of his fellow students, Gisbertus Carbo, to whom he had given the first skeleton he had articulated from bones he obtained by robbing the gallows.

PARIS

Medical School

In 1533, Vesalius entered the distinguished but extremely conservative medical school in Paris. He was to remain here for the next three years which in retrospect proved to be a crucial period in the development of his personality and critical approach to the study of human anatomy. What is known about Vesalius during this period has been derived from his own treatise and from the writings of others.

By the time Vesalius had reached the University of Paris, the works of both Galen and Hippocrates were available, having been translated directly from the Greek into Latin. Only a few years earlier, the learning of medicine was largely from the translations and commentaries of mediaeval and Moslem scholars. As remarked by Saunders and O'Malley (1982) "Physicians, seeing for the first time the works of Galen and Hippocrates stripped of their dross, believed that now they had captured the essence and spirit of the great classical authors and were at last about to enter a new Golden Age. As yet medicine had not developed a philosophy of progress but tended to look upon the present as inferior in knowledge and achievement to the past with the resultant enslavement to the literal word, and in particular to that of Galen."

Around this period, there were many eminent scholars who taught medicine at the University of Paris and attracted students from all parts of Europe. However, as far as anatomy was concerned, teaching continued to be based on the traditional enclave of Galenism and relatively few dissections were carried out, notwithstanding an appeal made by the Medical Faculty in 1526 to Parliament for more material to dissect.

Jacobus Sylvius (1478-1555) and Johannes Guenther of Andernach (1487-1574) must have profoundly influenced Vesalius. Mention is made also of Jean Vasse (1486-1550), Jean Fernel (1497-1558) and Oliverius in his work *Letter on the China Root.*

Jacobus Sylvius

Jacobus Sylvius was born at Amiens in 1478 (Fig. 87). He devoted his early years to classical philology and later became interested in anatomy after studying the Greek translations of Hippocrates and Galen. At a later stage in his life he decided to study medicine and graduated from the University of Montpellier with his doctorate degree when he was already fifty-one years old.

Sylvius collaborated with the noted surgeon, Jean Tagault, in carrying out dissections and increasing his knowledge of practical anatomy. He was industrious, eloquent and influential. He was an ardent follower of Galen, who even as the first Professor in France to have taught anatomy through human dissection, refused to accept any deviation from the writings of Galen even when his personal observations were at variance. In his opinion, it was more likely that the human body had changed from Galen's time.

We are indebted to Sylvius for introducing anatomical nomenclature, particularly with respect to the muscles. He described the technique of color injections, parts of the sphenoid bone, including the sphenoidal sinus, and the ventricles of the human brain. Some of

IACOBUS SYLVIUS
Regius Lutetia Professor

Figure 87. Jacobus Sylvius (1478-1555). (Wellcome Institute Library, London.)

these discoveries have been named for him.

Sylvius was apparently an oddity and a noted miser. Even the epitaph on his grave stated "Sylvius lies here, who never gave anything for nothing: Being dead, he even grieves that you read these lines for nothing" (Ball, 1910).

From Sylvius, Vesalius learned the technique of dissection as well as the traditional exposure to Galen's teachings. In the preface of his book, Vesalius was effluent in his praise for Sylvius although after the publication of Vesalius' book the two men fell apart and Sylvius filled with jealousy and rage became vindictive. In his works, Vesalius referred to Sylvius as "my erudite teacher Jacobus Sylvius" and as "the never-to-be-sufficiently-praised Jacobus Sylvius."

Later in his *Letter on the China Root* published in 1546, Vesalius clearly minimized the influence of Sylvius on his work by stating that:

> I don't know if Sylvius paid any attention to the remark in my books . . . that is that I have worked without help of a teacher, since perhaps he believes that I learned anatomy from him, he

who still maintains that Galen is always right. Sylvius, whom I shall respect as long as I live, always started the course by reading the books *On the Use of Parts* to us. . . . He brought nothing to the school except occasionally bits of dogs. . . . It happened one day that we showed him the valves of the orifice of the pulmonary artery and of the aorta, although he had informed us the day before that he could not find them. Since Sylvius omitted a chapter dealing with the vertebrae, as well as many others, during his so-called course — which he was then giving for the 35th time — and read nothing else anatomical except the books *On the Movement of Muscles* in which he everywhere agreed with Galen, it is not astonishing that I write that I have studied without the aid of a teacher (O'Malley, 1964).

The book of Sylvius' *In Hippocrates et Galeni Physiologiae Partem Anatomicam Isagoge* was published in 1555 but apparently it was completed in 1542. The work was essentially a systematic account of anatomy based on the teachings of Galen but included some personal observations as a result of human dissections. Sylvius was greatly influenced by the work of Niccolo Massa, *Liber introductorius anatomiae*, which although dated 1559 on the title page showed 1536 on the colophon.

Johannes Guenther

Johannes Guenther (1487-1574) was a Greek scholar who later taught anatomy at the University of Paris having obtained his baccalaureate degree in medicine in 1528 (Fig. 88). He translated many of the Greek scholars of antiquity, including the works of Galen which he also zealously promoted. Guenther's treatise *Institutiones Anatomicae* which appeared in 1536, was essentially a Galenic work in four parts. It followed a practical format of dissecting the organs that would decay first and others later.

According to Vesalius himself, the entire dissection and anatomical demonstrations were carried out over a period of three days and consisted of "for, except for eight abdominal muscles shamefully mangled and in the wrong order, no other muscle or any bone, and much less an accurate series of the nerves, veins or arteries was ever demonstrated to me by any." In his book, Guenther acknowledged the discoveries Vesalius made in regard to the spermatic vessels and later wrote that "when I presided at a public dissection at Paris . . . Andreas Vesalius assisted me" (O'Malley, 1964).

ΟΎΔΕΝ ΤΗΣ ΑΡΕΤΗΣ ΤΙΜΙΩΤΕΡΟΝ.

Figure 88. Johannes Guenther (1478-1574). (Wellcome Institute Library, London.)

Guenther was a sound but undistinguished anatomist who was affectionately regarded by his students. That he did not dissect a human body himself is apparent from Vesalius' remark that "I reverence him on many counts, and in my published writings I have honoured him as my teacher, but I wish there may be inflicted on my body, one for one, as many strokes as I have ever seen him attempt to make incisions in the bodies of men or beast, except at the dinner table." This should not come as a surprise in view of the fact that "the lecturers are perched up in a pulpit like jackdaws, and arrogantly prat about things they have never tried, but have committed to memory from the books of others, or placed in written form before their eyes. The dissectors are so ignorant of language that they are unable to explain the dissections to the onlookers, and merely botch what they are supposed to exhibit in accordance with instructions of the physician, who never applied his hand to the dissection, but contemptuously steers the ship out of manual, as the saying goes. Thus, everything is wrongly taught, days are wasted in absurd questions, and in the confusion less is offered to the onlooker than a butcher in his stall could teach a doctor" (O'Malley, 1964).

Guenther spent his last years travelling throughout Germany and Italy. Ferdinand I in 1562 conferred upon him the highest honours of nobility. He died in 1574.

Vesalius studied human bones obtained from the Cemetery of the Innocents and from Montfaucon where the bodies of executed criminals were left on the gallows. To his professors and fellow students he demonstrated a thorough knowledge of different parts of the skeleton which he was able to identify even blindfolded. In Paris, too, he was able to participate in human dissections of a "prostitute of fine figure and in the prime of life who had been hanged." Apparently this was the dissection to which Guenther had referred as "once in Paris when I presided over the public dissection of a female cadaver, Andreas Vesalius assisted me." He also dissected animals, including the dog, for the benefit of the class and remarked that "never would I have been able to accomplish my purpose in Paris, if I had not taken the work into my own hands."

Jean Fernel

Jean Fernel (1497-1558) was born near Amiens in France (Fig. 89). In 1538 Fernel began to write his *Universa Medicina* which was published in 1554. The first of the seven books, consisting of twenty chapters, was devoted to human anatomy. Although lacking illustrations it was a concise and original introduction to functional anatomy. Fernel was of the opinion that students should learn their anatomy through actual dissection rather than wasting their time looking at pictures in a book. He believed that sensory nerves originated from the brain, which was the seat of the soul, and motor nerves from the membrane. The discovery of the central canal of the spinal cord, *medulla spinalis cava est* is attributed to him (Sherrington, 1946).

LOUVAIN

In 1536 Vesalius returned to Louvain from Paris on account of

Figure 89. Jean Fernel (1497-1558). (With permission, from Major 1954).

the outbreak of the Franco-German war and the invasion of Pro-
vence by Charles V. As in Paris he continued his anatomical studies
by securing bones from the cemeteries and gallows. With the help of
Regnier Gemma (1508-1555), the well-known mathematician and
physician, he was able to obtain for the first time a complete human
skeleton, and the circumstances leading to this auspicious event can
best be appreciated from Vesalius' own words:

> While out walking, looking for bones in the place where on the
> country highways eventually, to the great convenience of stu-
> dents, all those who have been executed are customarily placed, I
> happened upon a dried cadaver. . . . The bones were entirely
> bare, held together by the ligaments alone, and only the origin
> and insertion of the muscles were preserved. . . . With the help of
> Gemma, I climbed the stake and pulled off the femur from the
> hip bone. While tugging at the specimen, the scapulae together
> with the arms and hands also followed, although the fingers of
> one hand, both patellae and one foot were missing. After I had
> brought the legs and arms home in secret and successive trips,
> leaving the head behind with the entire trunk of the body, I al-
> lowed myself to be out of the city in the evening in order to obtain

the thorax which was firmly held by a chain. I was burning with so great a desire . . . that I was not afraid to snatch in the middle of the night what I so longed for. . . . The next day I transported the bones home piecemeal through another gate of the city . . . and constructed that skeleton which is preserved at Louvain in the home of my very dear friend Gisbertus Carbo (Saunders and O'Malley, 1982).

Vesalius' skill in dissection was recognized and in 1537 he conducted one of the first human dissections to be held in the city for eighteen years.

Vesalius must have made a tremendous impression on his colleagues because not only did he dissect but he lectured at the same time. In February 1537, he published his first work *Paraphrase on the Ninth Book of Rhazes* which was submitted to the University of Louvain for the Baccalaureate (Candidate in Medicine) thesis. The book was printed by his friend Rutger Resch and marred by the poor production as a result of the old and worn types, as well as typographical errors (Fig. 90). A much improved edition appeared a month

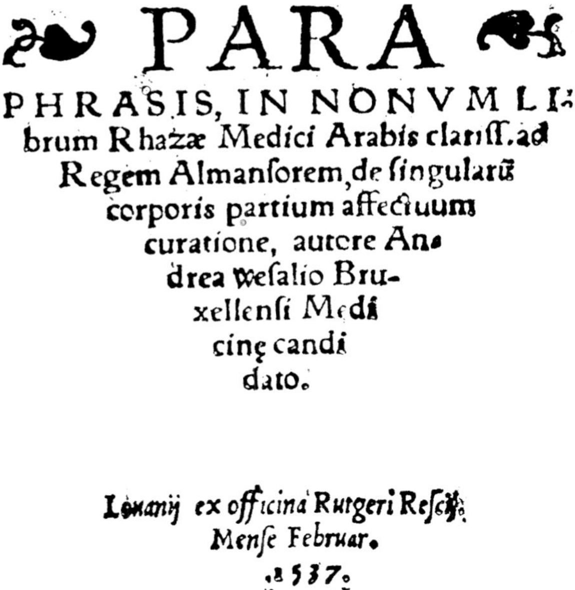

> ## PARA
> PHRASIS, IN NONVM LI-
> brum Rhazæ Medici Arabis clariff. ad
> Regem Almanforem, de fingularũ
> corporis partium affectuum
> curatione, autore An-
> drea Wefalio Bru-
> xellenfi Medi
> cine candi
> dato.
>
> Louanij ex officina Rutgeri Refch.
> Menfe Februar.
> 1537.

Figure 90. Title-page of the rare first edition of Vesalius' inaugural dissertation "Paraphrase of the ninth book of Rhazes," Louvain, February 1537. (From Cushing 1943; with kind permission of the Blackie Publishing Group, Glasgow.)

later in Basel. It was a free translation of the work of the Caliph Al-Mansur which covered all aspects of the healing art in ten books. In his own words, Vesalius' intention was to "revise the translation of Rhazes, both in order to liberate other candidates of medicine from immense labour, and to enable the author himself to reach men's hands henceforth cleansed of all barbarian names of medicaments unrecognizable to Latin ears, and with his entire mode of expression changed for the better; so that what was heretofore squalid and coarse and too obscure to be intelligible will now be brightened as far as possible, and will require the least exertion from the reader" (Cushing, 1943).

UNIVERSITY OF PADUA

In 1537 at the age of twenty-two, Vesalius arrived in Venice which at that time was a prosperous and enlightened city. Here he continued his pursuit of human anatomy encouraged by Ignatius Loyola and his fellow Jesuits. The Theatine monks were concerned with caring for the sick and recognized the need for anatomical knowledge. It was in Venice too that he met Jan Stephan Van Calcar, his countryman and student of Titian.

On December 5, 1537 Vesalius obtained his Doctor of Medicine degree from the University of Padua. The city of Padua was under the control of Venice and the University, which was established in 1222, rapidly evolved into one of the great intellectual centres of Europe. Its medical school was to have profoundly influenced the scientific renaissance. On the following day the Senate of Venice appointed Vesalius Professor of Surgery, with the responsibility also for Anatomy, at the University of Padua. On that very day the newly appointed Professor performed a dissection and delivered a lecture on the treatment of phlegmon. As "Lector Ostensor et Incisor," Vesalius was apparently the first person to have been appointed to a chair in Anatomy and also the first person to have received a salary for teaching the subject at any University.

Vesalius devoted himself enthusiastically to the work of teaching and writing, dissecting with his own hands as he taught his students. As in Louvain, Vesalius carried out the dissections himself and demonstrated the anatomy to his students. He employed his students as dissectors rather than the ignorant barbers. The dissections were carried out during the winter months because of putrefaction.

In contrast to the course given by Mondino, Vesalius dissected the entire day for a period of three weeks in order to demonstrate the anatomy of the human body. After a study of the skeleton, the muscles, blood vessels and nerves were demonstrated followed by the abdominal organs, chest and brain in another cadaver. He drew large diagrams in order to simplify the task for his students. Dissection of domestic animals and comparative anatomy were also carried out but only in order to elucidate features of human anatomy.

At first Vesalius lectured and dissected in one of the private houses, the Casa de'valli near the Ponte della paglia, in the city, but later he conducted his dissections and demonstrations in an aula built of wood and with a capacity for 500 persons. In the center of the room was a table on which bones and joints were kept and at the end of the table was an articulated skeleton to which he constantly referred.

Vesalius was supremely successful as a teacher because of his skill as a dissector and his enthusiastic expositions and demonstrations which he personally carried out. Indeed, his greatest innovation as an anatomist was that he dissected himself and lectured at the same time. His course was popular and attended not only by the medical students, but by many members of the university, government officials, clergy, artists, and prominent citizens of the city.

TABULAE ANATOMICAE

Within the first five months of his professorship, Vesalius articulated a human skeleton for use by his students in the classroom, published a set of six anatomical plates, the *Tabulae Anatomicae*, in April 1538, and in the same month issued a new edition of the *Institutiones Anatomicae*, a work by his former teacher, Guenther, for the benefit of his students. The *Tabulae Anatomicae* consisted of six loose leaf plates (*Fliegende Blätter*). The plates were more than sixteen inches in length and cut in wood. The drawings were realistic and from the actual dissections. The first three drawings dealt with the vascular system and were done by Vesalius himself. The other three, showing the skeleton, were drawn by his friend Van Calcar and considered to be more accurate than those of Berengario (1522) and of Balamio (1535).

The success of Vesalius' first published anatomical work was evident from the fact that the anatomical plates were widely plagiarized in Germany and France (Cushing, 1943). Vesalius was so bothered that in a letter to his publisher, Oporinus, he bitterly complained:

> But these privileges are often not worth the paper they cover, as I know but too well, from what happened to my *Anatomical Tables* published at Venice three years ago, the value of the decrees of sovereigns issued to printers and booksellers who swarm everywhere and have marred my works in putting them forth under pompous titles. In the edition of Augsburg the letter which I have written to Narcissus Vertunus, that model of physicians of our time, and first physician to the Emperor and king of Naples, has been suppressed and replaced by the preface of a German babbler who, unworthily decrying Avicenna and the Arabic authors, ranks me with the abridged Galens and, to cheat the buyer, pretends that I have put into six tables what in Galen fills thirty books.
>
> Moreover, he affirms that in his German translation he has used Greek and Latin terms, whereas he has not only suppressed these but has likewise omitted all that he could not translate and nearly all that gave value to my *Tables*; to say nothing of his wretched copies of the Venetian prints. He who has pirated them at Cöln has done still worse than his fellow Augsburg; and although some anonymous scribe has written in praise of the printer that his plates are better than mine, they are in truth merely clumsy copies of rough sketches of my own which I had communicated to some of my friends during the progress of the work. At Paris they have copied the three first plates very well, but the others they have omitted, perhaps because they were difficult to engrave, though it was these first three which students could have best dispensed with. Worst of all, so far as science is concerned, at Strasburg, another plagiarist, whom Fuchs greatly blamed, thought fit to contract the size of the plates, which can hardly be too large, to daub them with ugly colouring, and to surround them with a text borrowed from that of the Augsburg edition, but put forth as his own. Envying this man's fame, another is publishing a book, compiled from all quarters and illustrated with prints taken from books printed at Marburg and Frankfort. Indeed I would gladly put up with and even welcome [the cost entailed on me by] the divine and happy Italian wits [whose works] deserve an appreciation from the doctors in Germany different from that which they have received from those who know them only as reproduced by

the miserable slaves of sordid printers who, seizing upon any writing from which a little profit can be wrung, abridge it without discernment, or alter it, or merely copy it, and publish it under their own names as something new, and as if it were not protected by any privilege (Cushing, 1943).

It is a rare work of which only three complete copies are known to have survived (Library of San Mario, Venice; Hunterian Library, University of Glasgow; Prague) notwithstanding several Galenic misrepresentations such as the five-lobed liver, the *rete mirabile* at the base of the brain, the work was an immediate success because of "the large size of the plates, their fidelity to nature and the skill with which they were cut in wood" (Ball, 1910).

The *Tabulae Anatomicae* (also known as *Tabulae Sex*) were used by medical students at that time in conjunction with Guenther's *Institutiones Anatomicae*. The latter work was essentially a synopsis of Galen's concepts of anatomical structures and functions. In the first edition, Guenther remarked on his former pupil as "Andreas Wesalius — a young man of great promise who possesses an extraordinary knowledge of medicine, learned in both (classical) tongues and skilled dissection."

VESALIUS AND GALENISM

Vesalius emphasized that the only true source of anatomical knowledge was the human body itself which often proved to be different from that described by Galen. In 1540, Vesalius made a dramatic demonstration in Bologna of the skeletons of a man and an ape and demonstrated more than two hundred differences where Galen was mistaken with respect to the human body but not to that of the ape. Nevertheless, he made three major contributions (1. *Dissection of the Nerves*, 2. *Dissection of the Veins and Arteries*, 3. *Anatomical Administrations*) to a new and complete edition of Galen's work published by Lucantonio Giunta assisted by Augustinus Sadaldinus. This expansive work in seven volumes, *Galeni omnia opera*, appeared between 1541 and 1542. It was edited by the famous Joannes Baptista Montanus of Verona, Vesalius' former teacher and friend. According to Cushing (1943), the book was immensely popular and as many as fourteen editions were printed in Venice, Basel and Lyon by the publishing firm of Froben within a period of eighty years.

A Galenist at first, not unlike his contemporaries, Vesalius was later to remark that "it is quite clear to us, from the revival of the art of dissection, from a painstaking perusal of the works of Galen, and from a restoration of them in several places, of which we have no reason to be ashamed, that Galen himself never dissected a human body lately dead." Vesalius saw the immediate need for a book on anatomy based upon actual dissection and original observations. He relentlessly pursued this with single-minded zeal for the next few years. The work, however outrageous it might have seemed to the Galenists at that time, was quickly recognized as a brilliant exposition of the true structure of the human body.

CHAPTER THIRTEEN

DE HUMANI CORPORIS FABRICA

BEGINNING OF MODERN ANATOMY

THE publication of Vesalius' *Fabrica* ushered a new era in the history of medicine and marked the beginning of modern anatomy. The work emanated from an analytical mind who knew that in order to describe the true structure of the human body he had to first dissect it. Vesalius' observations were not always in agreement with the established teachings of Galen which were slavishly accepted for more than twelve centuries. Nevertheless, he had the courage to describe what he saw, for in science lies only truth.

In order to pursue his work, Vesalius was faced not only with scientific matters, but he also had to surmount sensitive moral and philosophical problems. During the Middle Ages, the art of healing was highly regarded but anatomy, which was concerned with the dead body, occupied a very low position. The dissection of the dead and decaying human body stood in stark contrast to the classical humanism of the Renaissance. It was an era, too, when spiritual values were held supremely over material things (Richardson, 1885; Roth, 1892; Singer, 1943; Cassirer, 1943).

Vesalius' work was profoundly original. He was not only "engaged in a continuous struggle against philosophical authorities," but he also "denied and defied the scholastic tradition." Science and philosophy were to experience profound changes with human anatomy evolving into a pure empirical branch of science, having survived its intellectual crises (Cassirer, 1943).

WRITING AND PUBLICATION
OF VESALIUS' FABRICA

Vesalius probably started work on his book in the winter of 1539 and by the summer of 1542 it was completed. The first indication of his plans for such a work is to be found in the last paragraph of his dedicatory preface to Narcissus Parthenopeus Vertunus, who was a physician to Charles V. Vesalius implored his patron to "accept therefore, most honored sir, this trifling gift on paper with that graciousness with which you once received me when you declared yourself, with many tokens of benevolence, as being particularly well disposed towards me. If I shall find this work acceptable to you and to students, some day I hope to add something greater. Farewell. Padua, the First day of April in the year of our Salvation 1538" (Saunders and O'Malley, 1982).

A year later in his *Venesection Letter* he remarked that "we have now also finished the two plates on the nerves; in the first, the seven pairs of cranial nerves have been drawn and in the other, all the small branches of the dorsal medulla expressed. I consider that these must be kept until we undertake the plates on the muscles and all the internal parts." This would indicate that the book was beyond the planning stage. In regard to the drawings he remarked that "if the opportunity of bodies offers, and Jan Stefan (Van Kalkar), outstanding artist of our age, does not refuse his services, I shall by no means evade that labour" (Saunders and O'Malley, 1982).

In August 1542, Vesalius and his friend Nicolaus Stopius packed the manuscript and engraved wood blocks which were sent to the publishers, Robert Winter and Johannes Oporinus, in Basel. The journey from Venice across the Alps was accomplished on mule back. Accompanying the manuscript and illustrations were the most precise and detailed instructions probably ever sent by an author to any publisher for the printing of a book.

> "You will shortly receive . . . together with this letter, the plates engraved for my *De humani corporis fabrica* and for its *Epitome*. I hope they will reach Basel safely and intact; for I have carefully packed them with the help of the engraver and of Nicolaus Stopius, the faithful manager of the firm of Bomberg and a young man well versed in humanities, so that they may nowhere rub against one another or receive any other damage in transit. We have inserted separately and in sequence between the plates,

schemata ["exemplar"], together with the letterpress belonging to
individual figures, to indicate where each is to go, so that their
order and disposition might offer no difficulties to you or to your
assistants and that the figures might not be printed out of sequence. You will quickly see from the directions [?] where the nature of the characters should be changed, since I have separated
by means of ruled lines that part of the letterpress which describes
the history of the organs in the text divided into chapters, from
that which serves as an explanation to the characters engraved on
the plates, calling the latter for this reason the Index of the plates
and their characters. In the continuous text, uninterrupted by
references to figures, please use the types called in printing-
offices "superlinear," to indicate the annotations which I have
added in the inner margin, not with so much industry as with
great labour and tediousness, so that they might serve the reader
as a commentary showing on what plate the part discussed may
be seen; the annotations on the outer margin, on the other hand,
are meant to serve as a summary of the text. From the inner annotations, in order to avoid prolixity, I have omitted the number
of the chapter wherever the plate referred to is attached to that
chapter; where this is not the case, I have indicated the number
of the chapter also. The same applies to the number of the 'book.'
You will find it abundantly explained, in the titles of the 'books' or
in the indices of the characters, why I have assigned the plates to
this or that location. We have engraved on the plates the characters, of the sort always used in printing-offices, to indicate individual parts of each drawing; first we used capital letters, then
other Latin letters, then lower-case Greek letters, then some capital [Greek] letters such as are not identical with [capital] Latin
characters. Where all these were not sufficient, we have used
numbers. Where a character has but one explanation, it stands in
the margin all by itself; where it does not have it, it is followed in
the margin by a period, so as to make the proper relation clear to
the reader. I have already written to you, in greater detail, of my
reason for all this, and why I thought that the index of characters
[his term for explanation of the plates] should not be confused
with the description of the parts; now, however, I exhort you
most earnestly that everything should be executed clearly and
speedily, and that you should acquit yourself satisfactorily with
regard to my efforts and do justice to the expectations of everyone respecting your printing-office, now set up for the first time
for the greater convenience of students and under the happy aus-
pices of the Muses. The utmost in care should be applied to the
impression of the plates, so that they may be executed not ama-
teurishly [*vulgariter*] in the manner of elementary school-books
[*scholastice*] nor in simple lines; in no place should the precise

impression of the plate be neglected, except perhaps where the drawing is complemented by [verbal] description. And though in this regard your judgement is so good that I can expect the best from your industry and assiduity, there is one particular request I must make of you, that in producing this work you imitate as closely as possible every mark which you will find made by the engraver, in following his copy, and revealed on the wood-blocks; and thus no character, no matter how much in shadow, will escape the notice of an observant and careful reader. Let also what I consider most skillful and most pleasant to the eye, the thickness of the lines in some parts be softened by elegant shading. . . . I shall take care to come to you soon and to remain at Basel, if not through the entire printing period then at least for some time, and I shall bring with me a copy of the decree of the Venetian Senate cautioning everyone not to print any of these plates without my consent. My mother sends you from Brussels an imperial licence, since you usually have one such in all the books which you are the first to print; I have procured it a short while ago, but have not yet succeeded in renewing it so as to make it valid for several years longer. The French royal licence will be obtained for me by the French ambassador in Venice, the Count of Montpellier. I am but little troubled on this account, so long as I do not have to fill a whole page with the transcript of official documents (From Cushing, 1943).

In January 1543, Vesalius arrived in Basel and probably supervised the production of the book. The high regard of Johannes Oporinus (1507-1568) for Vesalius and his work is clear from his remarks at the beginning of the book.

We have received from Andreas Vesalius, now in Italy, the following letter sent to us, together with the plates prepared for this work *De humani corporis fabrica* and for its *Epitome*. It seems to us that this letter contains many things necessary to admonish the reader at the outset, and of significance also to printers, particularly since they esteem but slightly even decrees of princes and are accustomed to pervert such things as serve to spread literary matters among the public. We have deemed it worth the trouble, therefore, to communicate it to the candid readers precisely as it has been sent to us (Cushing, 1943).

Oporinus was a scholar as well as a publisher. He was Professor of Latin and Greek at the University and was familiar with medicine because he had started his career with Paracelsus. At that time it would have taken someone of more than remarkable courage to publish such an enormous, heavily illustrated and controversial manuscript. However, the success of Vesalius' book was to have added to Oporinus' reputation, but later he was imprisoned for publishing a Latin translation of the Quran (Rollins, 1943).

JAN CALCAR AND OTHER ARTISTS

With respect to the magnificent illustrations, it will perhaps never be resolved whether they were all drawn by Jan Calcar (Feyfer, 1933). The work has been attributed to one or more of several artists, including Jan Calcar, Titian and others of his school, Vesalius and even Leonardo da Vinci. Strong evidence in favour of Jan Calcar for many of the illustrations, if not all, was presented by Cushing (1943). He referred to the bareheaded young man in the front row with the open sketch book having the initials S.C. on the cover. For reasons discussed by the Vesalian scholars Saunders and O'Malley (1982), the conclusion was reached that the drawings for both the *Fabrica* and its accompanying *Epitome* were done by students of Titian. Not only Jan Calcar but several other artists worked under the supervision of Vesalius. It is undeniable that Vesalius himself must have drawn some of the sketches.

THE WOOD BLOCKS

The wood blocks were cut on flat boards of apple, pear, beech, or sycamore, but the person or persons responsible for this impressive craftsmanship is unknown. The wood cuts have obviously been made with great care because they illustrate the text with great fidelity (Rollins, 1943).

The long history of these wood blocks has been traced by both Cushing (1943) and Saunders and O'Malley (1982). After Oporinus' death in 1568 the publishing business passed on to the Forbens. The business was closed in 1603 and the wood blocks came into the possession of Andreas Maschenbauer who was a printer and publisher in Augsburg. He selected nineteen of the wood blocks, twelve of which were used exclusively in the *Fabrica* and four in the *Epitome*, for a work on surface anatomy which was first published in 1706. A different title page, with five skulls, and a new dedication to his patron, Gottfried Amman, were incorporated. Surprisingly, on the title page he added the name of Titian as the artist responsible for the illustrations. The second edition of this book was published in 1723 using the same wood cuts and an introductory note to the reader.

The wood blocks were mentioned again sixty years later in the preface of a condensed edition of the *Fabrica* that was published in 1783 by the Bavarian anatomist and surgeon, Leveling of Ingolstadt. From the preface of this book we learn that the plates of the *Fabrica* and *Epitome* were purchased by the Bavarian physician von Woltter with the intention of publishing them together with German commentaries. The task was later entrusted to Leveling. He presented evidence to show that the wood blocks were the original but believed that some of the smaller ones might have been lost and were replaced.

In 1800, the blocks were moved to Landshut on the Isar and were then transferred twenty-six years later to the University of Munich library. Here they were stored away to be rediscovered in 1893 by the University librarian, Dr. Hans Schnorr von Carolsfeld. Professor Roth studied the 159 blocks that were recovered and published a detailed account in 1895 (Roth, 1895).

The New York Academy of Medicine, in collaboration with the University of Munich, published in 1934 the *Icones Anatomicae* which contains the re-engraved front-piece of the 1555 edition and 228 illustrations. Fifty of the blocks were missing and the portrait of Vesalius was reproduced in facsimile. Sadly, these blocks which had survived more than four centuries were destroyed as a result of bombing during the Second World War.

OUTLINE OF THE WORK

Vesalius' book appeared in folio size ($16'' \times 11''$) and consisted of 659 numbered pages with a magnificent wood-cut title page (Fig. 91). and a portrait of the author (Fig. 92) by Jan Calcar. There are 277 plates, both full page and smaller sizes. The book is divided into seven sections in accordance with Galen's approach to dissection. The preface of six pages is directed to Emperor Charles V and is dated August 1, 1542. This is followed by Vesalius' letter of August 24 from Venice to the publisher Oporinus. The work is set in Roman type and there are many instances of typographical errors, as well as inconsistencies in the numbering of pages (Cushing, 1943).

The book is copiously illustrated (Figs. 93-99). Of particular interest is the continuous panorama of a background landscape that is

Figure 91. Title page of Vesalius' *De Humani Corporis Fabrica*, 1543 (Courtesy of Dr. F.D. Bertalanffy). A public dissection is in progress, conducted by Vesalius himself. This engaging wood-cut also adorned his *Epitome*. The significance of the frontispiece and the identity of individuals depicted have been the subject of much scholarly debate (see Speransky et al., 1983). In the second edition, published just two years later, a re-engraved title-page, with many minor modifications, was used. Vesalius is now more prominently represented, the nude figure holding the left column is clothed, a goat has been added to the dog in the right foreground, etc.

ANDREAE   VESALII.

Figure 92. The only known genuine portrait of Andreas Vesalius (from his *De Humani Corporis Fabrica*), a work of Jan Calcar done when Vesalius was 28 years old (Medical Library, University of Manitoba).

revealed when plates 1, 2, 6, 5, 4, and 3 are viewed adjacent to each other. This area has been identified as the Euganean Hills, just a few miles out of Padua.

The first book of 40 chapters deals with the skeletal system. In the second book muscles and ligaments are covered in sixty-two chapters. Fifteen chapters are devoted to the circulatory system in the third book and, in the fourth book, seventeen chapters are devoted to cerebral and peripheral nerves. In the fifth and sixth books,

Figure 93. Superficial muscles, viewed from the side. (All illustrations from Vesalius' work are reproduced here through the courtesy of the Medical Library, University of Manitoba.)

the abdominal and thoracic organs are described, respectively. The seventh book of nineteen chapters is devoted to the brain and organs of special senses. The final chapter deals with vivisection and animal experiments. It includes an illustration of a pig strapped to a board in preparation for experiments.

Embodied in the work was a scientific approach to the study of human anatomy. At the same time it boldly rejected Galenic tradition and authoritarianism. However, the physiological considerations of Vesalius were no more advanced than those of Galen (Roth, 1892; Lambert, 1936; Cushing, 1943).

PLAGIARISM, COPIES AND OTHER EDITIONS

After the publication of Vesalius' *Fabrica*, no other noteworthy anatomical book was to appear for over a century. The popularity of

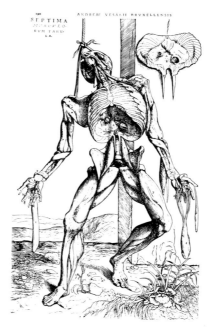

Figure 94. Another view of the muscles together with the diaphragm. For dissection, the cadaver was suspended in an upright position by a rope. Many muscles are readily recognizable and the openings for the inferior vena cava, esophagus and aorta in the diaphragm are precisely located.

the work can be judged from the extent that it was copied and plagiarized. That it did not meet the approval of the Galenists was to be anticipated because it amounted to a stinging repudiation of Galen and his teachings. About this, Vesalius remarked that "Many persons are hostile to me because in my writings I seem to hold in contempt the authority of Galen, the prince of physicians and preceptor of us all, because I do not agree indiscriminately with all his opinions, and especially because I have demonstrated that some errors are discernible in his books" (Cushing, 1943).

The most venomous of attacks came from Sylvius, his former teacher in Paris; Johannes Dryander in Marburg; Juan Valverde di Hamusco, a Spanish anatomist who had studied in Padua; Realdus Columbus, Vesalius' former assistant at Padua; Franciscus Puteus, who had studied medicine in Pavia and Bologna; and later by Bartholomeus Eustachius in Rome (Lambert, 1936; Castiglioni,

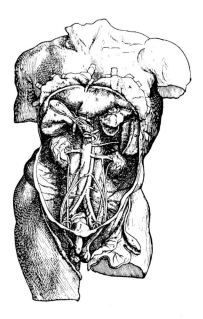

Figure 95. Dissection of the abdominal cavity. The intestine has been removed to reveal structures on the posterior abdominal wall. Observe the position (?) of the kidneys, distribution of blood-vessels and nerves, and the opened left scrotal sac.

1943; Cushing, 1943).

In the aftermath of Vesalius' work, the most important contribution to Renaissance anatomy was Gabriele Fallopius' (1523-1562) *Observationes Anatomicae* which was published in Venice in 1561 (Fig. 100). In the preface, Fallopius praised Vesalius for his work but at the same time was not uncritical of some errors. Fallopius, himself Professor of Anatomy, Surgery and Botany at the University in Padua, was a brilliant teacher and physician who recognized Vesalius as "the prince of anatomists, an admirable physician and a divine teacher." He, himself, is credited with noteworthy discoveries including the tubes that are named after him, the canal through which the facial nerve passes, the round and oval windows in the ear, and the communication between the tympanic cavity and mastoid ear cells.

Vesalius responded to the criticisms of Fallopius in a letter from

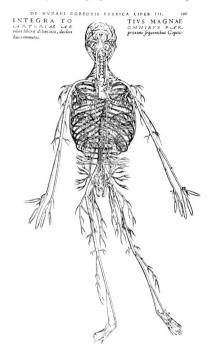

Figure 96. The entire arterial system. This illustration is a composite drawing of the arrangement of the vessels, derived largely from the dissection of animals.

Madrid in 1561. Fallopius died before receiving the letter because the emissary was delayed in Spain. It was subsequently published by Francesco Senese in Venice in 1564 after the death of Vesalius, with the title *Andreae Vesalii Anatomicarum Gabrielis Falopii Observationum Examen* (Fig. 101).

The second edition of the *Fabrica* was published by Oporinus in 1555. It was a more lavish volume with several changes, including a new wood-cut title page. By this time, Vesalius' work had spread throughout Europe as a result of re-engraved copies and plagiarism. In most cases the copied illustrations lacked the visual expressiveness that was so compellingly evident in the original 1543 edition of the work. Vesalius, himself, complained bitterly and in a letter to his publisher remarked "But who, I ask, can feel the faintest inclination to publish the results of his night long studies, when there are people everywhere guilefully plotting to destroy the works of others? A case in point exists at present in England, where the figures of my *Epitome*

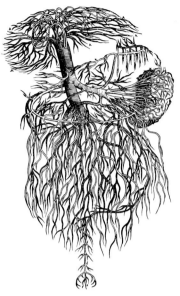

Figure 97. Representation of the portal system. There are many misconceptions, such as the division of the portal vein into five branches for the five lobes of the liver.

have been copied very poorly and without skill in drawing — although not without expense to whoever will have to pay for them. And, indeed, I would be ashamed to have anyone think that I had published these illustrations in such form" (Cushing, 1943).

Vesalius was undoubtedly referring to the enterprising Thomas Geminus, an Italian and engraver himself, who migrated to England and subsequently published in 1545 a *Compendiosa Totius Anatomie Delineatio, Aere Exarata* which consisted of forty pages of copper plates engraved from the work of Vesalius. An identical English version of the book appeared in London a year later. Geminus' books were the first with copper plates to be published in England. A second edition of the English translation with a dedication to Queen Elizabeth I appeared in 1559. The engravings attracted much attention and were copied in Germany.

In 1551 Jacob Bauman issued his *Anatomia Deudsch*. It contained forty copper engravings of poor quality that were copied from the Geminus plates. The legends and explanations were in German. Other editions of this work appeared in 1551 and 1575. Nothing new

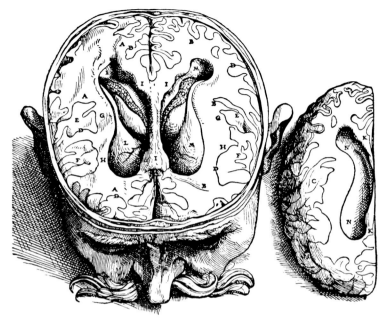

Figure 98. Horizontal section of the brain. The right and left cerebral hemispheres, as well as the lateral ventricles, corpus callosum, and the choroid plexus can be seen. One can also delineate the gray and white matter.

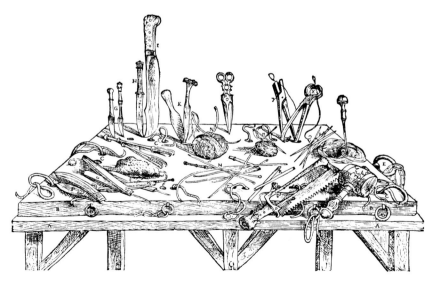

Figure 99. Dissecting instruments used by Vesalius.

GABRIELIS FALLOPPII
MEDICI MVTINENSIS
OBSERVATIONES
Anatomicæ.

AD PETRVM MANNAM
medicum Cremonenfem.

Cum Priuilegio Summi Pontificis,
Regis Philippi, Senatusque
Veneti.

VENETIIS.
Apud Marcum Antonium Vlmum
M D L X I.

Figure 100. Title-page of Fallopius' *Observationes anatomicae* issued in 1561 which pleased Vesalius and resulted in the *Examen* on Fallopius, Vesalius' final publication. (From Cushing 1943; with kind permission of the Blackie Publishing Group, Glasgow.)

was added but, nevertheless, the book remained popular in Germany for over two decades. It provided the basis for another anatomical text that was subsequently issued by Henricus Bottar, Dean of the Faculty of Medicine in Cologne. His *Epitome* was copied from the work of Geminus, but it had a new engraved portrait of Vesalius, taken from the 1543 edition of the *Fabrica*. According to Cushing (1943) the plates can be traced to those used by Bauman for his *Anatomia Deudsch*. An exact edition of this work reappeared in 1601 and in 1617 was reissued again in Amsterdam (Figs. 102-106).

EPILOGUE

Just twenty-nine years old and at the height of his fame, and

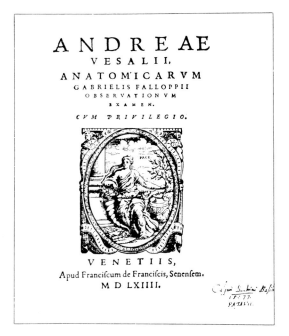

ANDREAE
VESALII,
ANATOMICARVM
GABRIELIS FALLOPPII
OBSERVATIONVM
EXAMEN.

CVM PRIVILEGIO.

VENETIIS,
Apud Francifcum de Francifcis, Senenfem.
M D LXIIII.

Figure 101. Title-page of Vesalius' *Examen* on Fallopius, 1564. (From Cushing 1943; with kind permission of the Blackie Publishing Group, Glasgow.)

only a few months after the publication of his great book, Vesalius succumbed to the harsh criticisms and abuse of his contemporaries by relinquishing his Chair at the University of Padua and burning all of his valuable unpublished papers. Of this act during December 1543, he was to later write in his *China Root* epistle that "as to my notes, which had grown into a huge volume, they were all destroyed by me; and on the same day there similarly perished the whole of my paraphrase of the ten books of Rhazes to King Almansor, . . . I was on the point of leaving Italy and going to Court; those physicians of whom you know had given the Emperor and the nobles a most unfavorable report of my books and of all that is published nowadays for the promotion of study; I therefore burnt all those works mentioned, thinking at the same time it would be an easy matter to abstain from writing for the future. I have since repented more than once for my impatience, and regretted that I did not take the advice of the friends who were then with me" (Cushing, 1943). Vesalius

Figure 102. Frontispiece of a rare copy of Vesalius' work. This edition was issued in Amsterdam in 1617 (Medical Library, University of Manitoba). The illustrations can be traced to Thomas Geminus, an engraver, who copied the Vesalian plates for his own book published in 1545.

became court physician to both Philip II and Charles V (*medicas familiaribus ordinaribus*) as a result of which he also participated in military expeditions.

For reasons unknown, Vesalius undertook a pilgrimage in 1564 to Jerusalem and on his return journey was shipwrecked on the small island of Zante where he died on October 15, 1564 (Wharton, 1902; Cullen, 1918; Lambert, 1936). There are many varied accounts of his death and none might be entirely accurate. Certain, however, is the scientific legacy he nurtured with obsessive devotion,

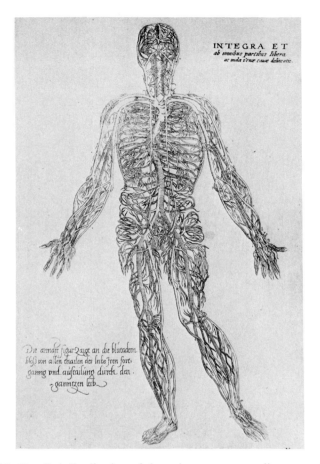

INTEGRA ET
ab omnibus partibus libera
ac nuda venæ cavæ delineatio.

Die anndt figur zaigt an die blutadern
blöß von allen thailen des leibs fren forts
gang vnd auftailung durch den
gantzen laib.

Figure 103. Detailed distribution of the veins — an outstanding copy, from the 1617 Amsterdam edition of Vesalius' work (Medical Library, University of Manitoba).

which made him "The first man of modern science." His pristinely preserved masterpiece, the *De Humani Corporis Fabrica*, is universally recognized as "one of the greatest treasures of western civilization and culture," which "established with startling suddenness the beginning of modern observational science and research" (Saunders and O'Malley, 1982).

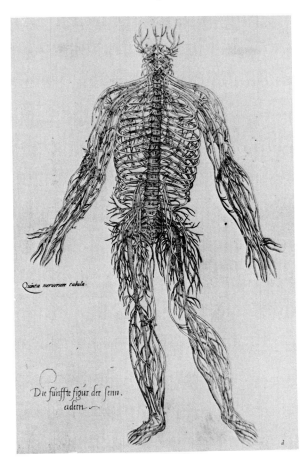

Quinta nervorum tabula.

Die fünffte figur der senn. adern.

Figure 104. Plagiarized copy (1617) of the original Vesalian plate showing the arrangement and distribution of (30 pairs of) spinal nerves (Medical Library, University of Manitoba).

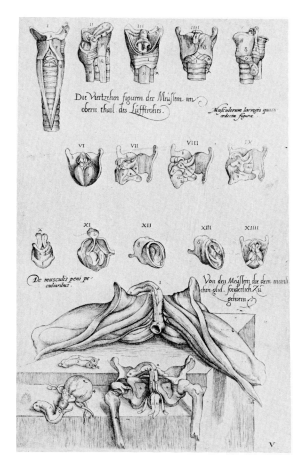

Figure 105. Re-engraved copy (1617) of Vesalian plates, showing the larynx and male external genitalia (Medical Library, University of Manitoba).

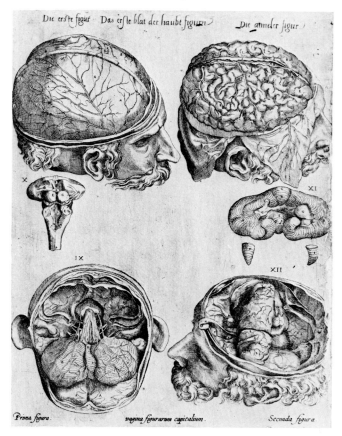

Figure 106. Composite re-engraved plate (1617) from the illustrations of Vesalius' Fabrica to demonstrate the dissection of the brain (Medical Library, University of Manitoba).

BIBLIOGRAPHY

Adams, F.: *The Genuine Works of Hippocrates. Translated from the Greek*. Baltimore, Williams & Wilkins Co., 1939.

Agrifoglio, L.: Anatomia e fisiologia del corpo umano in Cicerone. *Pag Stor Med,* 5:32, 1961.

Ahmed, R.U.: Status of anatomy and surgery in different civilizations and the contribution of Arabs in this field. In El-Gindy, A.R. and Hassan, H.M.Z. (Eds.): *Proc 2nd Int Conf Islamic Med*, Kuwait, 1982, pp.229-234.

Albar, M.A.: Embryology as revealed in the Quran and Hadith (Parts I & II). *IWMJ, 1*:46, 1983.

Andrews, C.A.R. and Hamilton-Patison, J.: *Mummies*. London, British Museum, 1978.

Arcieri, J.P.: *Why Alcmaeon of Croton is the Father of Experimental or Scientific Medicine*. New York, Alcmaeon Editions, 1970.

Aristotle: *Compleat Master Piece. Displaying the Secrets of Nature in the Generation of Man*. London, 1749.

Aristotle: *The Works of Aristotle, the Famous Philosopher, in Four Parts*. New England, 1831.

Aristotle: *On the Parts of Animals*: Translated, with Introduction and Notes by W. Ogle. London: Kegan Paul, Trench & Co., 1982.

Ashoor, A.A.: The history of Islamic medicine. *IWMJ, 1*:51, 1983.

Ashoor, A.A.: Islamic medicine: its influence on the Latin West. *IWMJ, 1*:44, 1984a.

Ashoor, A.A.: Muslim medical scholars and their works. *IWMJ, 1*:49, 1984b.

Ball, J.M.: *Andreas Vesalius. The Reformer of Anatomy*. St. Louis, Medical Science Press, 1910.

Belt, E.: *Leonardo the Anatomist*. Lawrence, University of Kansas Press, 1955.

Beltran, A.: *Rock Art of the Spanish Levant*. Translated by Margaret Brown: The Imprint of Man. Cambridge, University Press, 1982.

Bergsträsser, G.: Hunain ibn Ishaq über die syrischen und arabischen Galen-Übersetzungen. *Abh Kunde Morgenlandes, 17*:21, 1925.

Booth, N.B.: Empedocles' account of breathing. *J Hellen Stud, 80*:10, 1960.

Breasted, J.H.: *The Edwin Smith Surgical Papyrus*. Chicago, University of Chicago Press, 1930.

Browne, E.G.: *Arabian Medicine*. Cambridge, University Press, 1921.

Brunschwig, Hieronymus: *Buch der Cirurgia*. Strasbourg, 1497.

Burn, A.R.: *The Pelican History of Greece*. Middlesex, Penguin Books, 1982.

Campbell, D.: *Arabian Medicine and Its Influence on the Middle Ages*. London: Kegan Paul, Trench, Trübner and Co., Ltd., 1926.

Cappelletti, A.J.: Las doctrinas anatomo-fisiologicas de Diogenes de Apollonia. *Riv Storia Med, 18*:11, 1975.

Cassirer, E.A.: The place of Vesalius in the culture of the Renaissance. *Yale J Biol Med, 16*:109, 1943.

Castiglioni, A.: Aulus Cornelius Celsus as a historian of medicine. *Bull Hist Med, 8*:857, 1940.

Castiglioni, A.: *A History of Medicine*. Translated by E.B. Krumbhaar. New York, Alfred A. Knopf, 1941.

Castiglioni, A.: The attack of Franciscus Puteus on Andreas Vesalius and the defence by Gabriel Cuneus. *Yale J Biol Med, 16*:135, 1943.

Celsus: *De Medicina*. Translated by W.G. Spencer. 2 Volumes. London and New York, Loeb Classical Library, 1935.

Chadwick, J. and Mann, W.N.: *The Medical Works of Hippocrates*. Oxford, Blackwell Scientific Publications, 1950.

Chiera, E.: *They Wrote on Clay*. Chicago, University of Chicago Press, 1938.

Choulant, L.: *Geschichte und Bibliographie der Anatomischen Abbildung nach ihrer Beziehung auf Anatomische Wissenschaft und Bildene Kunst*. Leipzig, Rudolph Weigel, 1852.

Choulant, L.: *History and Bibliography of Anatomic Illustration*. Translated and edited with notes and a biography by Mortimer Frank. Chicago, The University of Chicago Press, 1920.

Clark, G.: *A History of the Royal College of Physicians of London*. Volume 1. Oxford, Clarendon Press, 1964.

Clark, K.: *A Catalogue of the Drawings of Leonardo da Vinci*. Cambridge, University Press, 1935.

Clark, K.: *The Drawings of Leonardo da Vinci in the Collection of Her Majesty the Queen at Windsor Castle*. Vols. 1-3. London, Phaidon Press Ltd., 1968.

Clarke, E.: Aristotelian concepts of the form and function of the brain. *Bull Hist Med, 37*:1, 1963.

Clarke, E. and Stannard, J.: Aristotle on the anatomy of the brain. *J Hist Med, 18*:130, 1963.

Cockburn, A. and Cockburn, E.: *Mummies, Disease and Ancient Cultures*. Cambridge, University Press, 1980.

Codellas, P.S.: Alcmaeon of Croton: his life, work, and fragments. *Proc Roy Soc Med, 25*:1041, 1932.

Cooper, S.: The medical school of Montpellier in the fourteenth century. *Ann Med Hist, 2*:164, 1930.

Corner, G.W.: *Anatomical Texts of the Earlier Middle Ages*. Washington, D.C., Carnegie Institution, 1927.

Corney, B.G.: Some physiological phantasies of third century repute. *Proc Roy Soc Med, 1*:217, 1914.

Cornford, F.M.: *Plato's Cosmology*. The Timaeus of Plato translated with a running commentary. London, Routledge and Kegan Paul, 1971.

Crummer, L.: Early anatomical fugitive sheets. *Ann Med Hist, 5*:189, 1923.

Crummer, L.: Further information on early anatomical fugitive sheets. *Ann Med Hist, 7*:1, 1925.

Cullen, G.M.: The passing of Vesalius. *Edin Med J, 13*:324, 1918.

Cullen, G.M.: Madrid to Zante. *Edin Med J, 13*:388, 1918.

Cushing, H.: *A Bio-Bibliography of Andreas Vesalius*. New York, Schuman's, 1943.

Dawson, W.R.: *ClioMedica: The Beginnings: Egypt and Assyria*. New York, P.B. Hoeber, 1930.

De Fanu, W.: A primitive anatomy: Johann Peyligk's "Compendiosa declaratio." Leipzig 1513. *Ann Roy Coll Surg Eng, 31*:115, 1962.

De Lint, J.G.: Beiträge zur Kenntnis der Anatomischen Namen im Älten Ägypten. *Sudhoff Arch 25*:382, 1932.

Dempster, W.T.: European anatomy before Vesalius. *Ann Med Hist, 6*:307 & 448, 1934.

Dhorme, E.: *Les Religions de Babylonie et d'Assyrie*. Paris, 1949.

Diller, H.: Die Lehre vom Blutkreislauf, eine verschlossene Entdeckung der Hippokratiker? *Sudhoff Arch, 31*:201, 1938.

Dobson, J.F.: Herophilus of Alexandria. *Proc Roy Soc Med, 18*:19, 1925.

Dobson, J.F.: Erasistratus. *Proc Roy Soc Med, 20*:825, 1927.

Dryander, J.: *Anatomia Mundini*, 1541.

Duckworth, W.L.H.: *Galen: On Anatomical Procedures. The Later Books*. (Translation.) Edited by M.C. Lyons and B. Towers. Cambridge, University Press, 1962.

Ebbell: *The Papyrus Ebers: The Greatest Egyptian Medical Document*. Copenhagen, Levin & Munksgaard, 1937.

Edelstein, L.: The development of Greek anatomy. *Bull Hist Med, 3*:235, 1935.

Edelstein, L.: The History of Anatomy in Antiquity. Temkin, O. and Temkin, C.L. (Eds.): *Ancient Medicine*. Baltimore, Johns Hopkins Press, 1967.

El-Gindy, A.R. and Hassan, H.M.Z.: *Proc 2nd Int Conf Islamic Med*. Kuwait: Kuwait Foundation for Advancement of Sciences, 1982.

Engelbach, R. and Derry, D.: Quoted in "Mummification." *Annales du Service des Antiquités de l'Egypte*, 41:236, 1942.

Erhard, H.: Alkmaion, der erste Experimentalbiologe. *Sudhoff Arch, 34*:77, 1941a.

Erhard, H.: Diogenes von Apollonia als Biologe. *Sudhoff Arch, 34*:335, 1941b.

Ferngren, G.B.: A Roman declamation on vivisection. *Trans Coll Physicians Phila, 4*:272, 1982.

Feyfer, F.M.G. de: Die Schriften des Andreas Vesalius. *Janus, 19*:435, 1914.

Feyfer, F.M.G. de: Jan Steven van Calcar (Joannes Stephanus), 1499-1546. *Nederl T Geneesk, 77*:3562, 1933.

Galen: *On Anatomical Procedures. The Later Books*. Translation by Duckworth, W.L.H.; Cambridge, University Press, 1962.

Galen: *On the Usefulness of the Parts of the Body. De usu partium*. Translation from the Greek with an Introduction and Commentary by M.T. May. Vols. 1 & 2.

New York, Cornell University Press, 1968.

Garrison, F.H.: Anatomical illustrations before and after Vesalius. *New York J Med, 101*:489, 1915.

Ghalioungui, P.: The West denies Ibn al-Nafis's contribution to the discovery of the circulation. In El-Gindy, A.R. and Hassan, H.M.Z. (Eds.): *Proc 2nd Int Conf Islamic Med*, Kuwait, 1982, pp. 299-304.

Gordon, B.L.: *Medieval and Renaissance Medicine*. New York, Philosophical Library, 1959.

Grapow, H.: Über die Anatomischen Kenntnisse der Altägyptischen Ärzte. Leipzig, J.C. Hinrichs, 1935.

Grayson, A.K.: Babylonia. In Cotterell, A. (Ed.): *The Encyclopedia of Ancient Civilizations*. London, Rainbird Publishing Group Limited, 1980, pp. 89-101.

Green, R.M.: *A Translation of Galen's Hygiene (De Sanitate Tuenda)*. Springfield: Charles C Thomas, 1951.

Green, R.M.: *Asclepiades: His Life and Writings*. New Haven, E. Licht, 1955.

Grene, M.: Aristotle and modern biology. *J Hist Ideas, 33*:395, 1972.

Grensemann, H.: *Der Arzt Polybos als Verfasser hippokratischer Schriften*. Mainz, Akad. Wissensch. Lit., 1968.

Gruner, O.C.: *A Treatise on the Canon of Medicine of Avicenna*. Incorporating a Translation of the First Book. London, Luzac & Co., 1930.

Gudger, E.W.: Pliny's "Historia naturalis": the most popular natural history ever published. *ISIS, 6*:269, 1923.

Harrington, J.: *The School of Salernum*. New York, Paul B. Hoeber, 1920.

Harris, C.R.S.: *The Heart and the Vascular System in Ancient Greek Medicine from Alcmaeon to Galen*. Oxford, Clarendon, 1973.

Harris, J.E. and Weeks, K.R.: *X-Raying the Pharaohs*. New York, Charles Scribner's Sons, 1973.

Hayek, S.: How Al-Zahrawi reached the Occident. *IWMJ, 1*:49, 1984.

Herrlinger, R.: Die didaktische originalität der anatomischen Zeichnungen Leonardos. *Anat Anz, 99*:366, 1953.

Herrlinger, R.: *History of Medical Illustration*. London, Pitman Medical & Scientific Publishing Co. Ltd., 1970.

Hoernle, A.F.R.: *Studies in the Medicine of Ancient India. Part I. Osteology or the Bones of the Human Body*. Oxford, Clarendon Press, 1907.

Hooper, A.: Further information on the prehistoric representations of human hands in the cave of Gargas. *Med Hist, 24*:214, 1980.

Huard, P. and Wong, M.: *Chinese Medicine*. New York, McGraw-Hill Book Company, 1968.

Hübotter, F.: *Die chinesische Medizin*. Leipzig: Verlag der Asia Major, 1929.

Hundt, M.: *Antropologium* . . . proprietatibus, de elementis, partibus, et membris humani corporis. 1501. (Copy in the British Library, London.)

Hyrtl, J. *Das Arabische und Hebräische in der Anatomie*. Wien, 1879.

Ilberg, J.: Über die Schriftstellerei des Klaudios Galenos. *Rhein Mus Phil, 44*:207, 1889.

Ilberg, J.: Über die Schriftstellerei des Klaudios Galenos. *Rhein Mus Phil, 47*:489, 1892.

Ilberg, J.: Über die Schriftstellerei des Klaudios Galenos. *Rhein Mus Phil, 51*:165, 1896.

Ilberg, J.: Über die Schriftstellerei des Klaudios Galenos. *Rhein Mus Phil, 52*:591, 1897.

Ilberg, J.: Wann ist Galenos geboren? *Sudhoff Arch, 23*:289, 1930.

Jablonski, W.: Die Theorie des Sehens im griechischen Altertume bis auf Aristoteles. *Sudhoff Arch, 23*(4):306, 1930.

Jaeger, W.: *Aristotle*. 2nd ed. London, Oxford University Press, 1948.

Janson, H.W.: *History of Art*. New York, Harry N. Abrams, Inc., 1977.

Janssens, P.A.: Medical views on prehistoric representations of human hands, *Med Hist, 1*:318, 1957.

Jastrow Jr., M.: The medicine of the Babylonians and Assyrians. *Proc Roy Soc Med, 7*:109, 1914.

Johanson, D.C. and Edey, M.A.: *Lucy: The Beginnings of Human Evolution*. New York, Simon & Schuster, 1981.

Johanson, D.C. and Edey, M.A.: *Lucy: The Beginnings of Human Kind*. New York, Warner Books, 1982.

Johanson, D.C. and White, T.D.: On the status of *Australopithecus afarensis*. *Science, 207*:1104, 1980.

Jones, W.H.S.: Hippocrates and the corpus Hippocraticum. *Proc Brit Acad, 31*:1, 1945.

Jouanna, J.: Le medecin Polybe est-il l'auteur de plusieurs ouvrages de la collection hippocratique. *Rev Etud grecq, 82*:393, 1969.

Jouanna, J.: Le medecin Polybe est-il l'auteur de plusieurs ouvrages de la collection hippocratique. *Rev Etud grecq, 82*:552, 1969.

Kapferer, R.: *Die anatomischen Schriften. Die Anatomie des Herz. Die Adern in der Hippokratischen Sammlung*. Stuttgart, Hippokrates Verlag, 1951.

Karim, M.A.: Contributions of Islamic medicine to anatomical sciences. In El-Gindy, A.R. and Hassan, H.M.Z. (Eds.): *Proc 2nd Int Conf Islamic Med*, Kuwait, 1982, pp. 196-200,.

Keele, K.D.: *Leonardo da Vinci on Movement of the Heart and Blood*. London, Harvey and Blythe Ltd., 1952.

Keele, K.D.: Three masters of experimental medicine — Erasistratus, Galen and Leonardo da Vinci. *Proc Roy Soc Med, 54*:577, 1961.

Keele, K.D.: Leonardo da Vinci's views on arteriosclerosis. *Med Hist, 17*:304, 1973.

Keele, K.D.: Leonardo da Vinci's "Anatomia Naturale." *Yale J Biol Med, 52*: 369, 1979.

Keele, K.D.: *Leonardo da Vinci's Elements of the Science of Man*. New York, Academic Press, Inc., 1983.

Keswani, N.H.: Susruta, the pioneer anatomist and the father of surgery. In Singhal, G.D. and Guru, L.V. (Eds.): *Anatomical and Obstetric Considerations in Ancient Indian Surgery*, Banaras, Banaras Hindu University Press, 1973.

Ketham, J.: Fasciculus Medicine. 1941. (Copy in the British Library, London.)

Ketham, J.: *The Fasciculo di medicina, Venice, 1493*. With an introduction, etc. by Charles Singer (Monumenta Medica, II), 2 vols. facsims. fol. Florence, R.

Lier, 1925.

Kevorkian, J.: *The Story of Dissection*. New York, Philosophical Library, 1959.

King, L.S.: Plato's concepts of medicine. *J Hist Med, 9*:38, 1954.

Klein-Franke, F.: *Vorlesungen über die Medizin im Islam*. Beiheft 23. Sudhoffs Archiv. Zeitschrift für Wissenschaftsgeschichte. Wiesbaden, Franz Steiner Verlag GMBH, 1982.

Koelbing, H.: Zur Sehtheorie im Altertum: Alkmeon und Aristoteles. *Gesnerus (Aarau), 25*:5, 1968.

Kraft, F.: Anaximandros. In: von Fassman et al. (Eds.), *Die Grossen der Weltgeschichte*,Vol. 1, Zurich, Kindler, 1971a.

Kraft, F.: Anaximandros und Hesiodos. Die Ursprünge rationaler griechischer Naturbetrachtung. *Sudhoff Arch, 55*:152, 1971b.

Kramer, S.N.: *Sumerian Mythology*. New York, Harper, 1961.

Kramer, S.N.: *The Sumerians: Their History, Culture and Character*. Chicago, Chicago Univ Press, 1963.

Kudlien, F.: Mondinos Standort innerhalb der Entwicklung der Anatomie. *Med Mschr, 18*:210, 1964.

Kühn, C.G.: *Claudii Galeni Opera omnia* (20 vols.). *Medicorum Graecorum Opera quae extant*. Leipzig, 1821-1833.

Lambert, S.W.: A reading from Andreae Vesalii, de Corporis Humani Fabrica Liber VII De Vivorum sectione nonnulla caput XIX. *Bull NY Acad Med, 12*:346, 1936a.

Lambert, S.W.: The physiology of Vesalius. *Bull NY Acad Med, 12*:387, 1936b.

Lassek, A.M.: *Human Dissection: Its Drama and Struggle*. Springfield, Charles C Thomas, 1958.

Leake, C.D.: *The Old Egyptian Medical Papyri*. Kansas, University of Kansas Press, 1952.

Leakey, R.E.: *Die Suchen nach dem Menschen. Wie wir wurden, was wir sind*. Umschau Verlag, Sigma Press und PR, Frankfurt (M), 1981.

Leca, A.-P.: *The Egyptian Way of Death: Mummies and the Cult of the Immortal*. New York, Doubleday, 1981.

Lehman, H.: Zu Constantinus Africanus. *Sudhoff Arch, 24*:263, 1931.

Leiser, G.: Medical education in Islamic lands from the seventh to the fourteenth century. *J Hist Med, 38*:48, 1983.

Leroi-Gourhan, A.: *The Dawn of European Art*. Translated by Sara Champion. Cambridge, University Press, 1982.

Lewin, R.: Fossil Lucy grows younger, again. *Science, 219*:43, 1983.

Lind, L.R.: *Studies in Pre-Vesalian Anatomy*. Philadelphia, The American Philosophical Society, 1975.

Lint, J.G. de: Fugitive anatomical sheets. *Janus, 28*:78, 1924.

Lipsett, W.G.: Celsus—first medical historian. *Hamdard med Dig, 5*:13, 1961.

Lloyd, G.E.R.: *Aristotle: The Growth and Structure of his Thought*. Cambridge University Press, 1968.

Lloyd, G.E.R.: Alcmaeon and the early history of dissection. *Sudhoff Arch, 59*:113, 1975a.

Lloyd, G.E.R.: The Hippocratic question. *Class Quart, 25*:171, 1975b.

Lloyd, G.E.R.: A note on Erasistratus of Ceos. *J Hellen Stud, 95*:172, 1975c.

Locy, W.A.: Anatomical illustrations before Vesalius. *J Morph, 22*:945, 1911.

Longrigs, J.: The "roots of all things". *ISIS, 67*:420, 1976.

Lonie, I.M.: Erasistratus, the Erasistrateans and Aristotle. *Bull Hist Med, 38*:426, 1964.

Lovejoy, C.O.: The origin of man. *Science, 211*:341, 1981.

Macalister, A.: The oldest anatomical memoranda extant. *J Anat, 32*:775, 1898.

MacKinney, L.C.: The beginnings of western scientific anatomy: new evidence and a revision in interpretation of Mondeville's role. *Med Hist, 6*:233, 1962.

MacKinney, L.: *Medical illustrations in medieval manuscripts.* Wellcome Historical Medical Library, London, 1965.

Magner, L.N.: *A History of the Life Sciences.* New York, Marcel Dekker, Inc., 1979.

Mallowan, M.E.L.: *Early Mesopotamia and Iran.* New York, McGraw-Hill, 1965.

Major, R.H.: *A History of Medicine.* Vol. 1. Springfield, Charles C Thomas, 1954.

Matsen, H.S.: *Allessandro Achillini and His Doctrine of "Universals" and "Transcendentals."* Ph.D. Diss., Columbia University, 1969. (Cited in Lind, 1975.)

Matthiae, P.: *Ebla: An Empire Rediscovered.* New York, Doubleday, 1981.

May, M.T.: Galen on human dissection. *J Hist Med, 13*:409, 1958.

May, M.T.: *Galen on the Usefulness of the Parts of the Body. De Usu Partium.* Vols. 1 & 2. Translated from the Greek with an introduction and commentary. New York, Cornell University Press, 1968.

McMurrich, J.P.: *Leonardo da Vinci, the Anatomist (1452-1514).* Baltimore, The Williams & Wilkins Co., 1930.

Meissner, B.: *Babylonien und Assyrien.* 2 Vols., Heidelberg, 1920-1925.

Meyerhoff, M.: New light on Hunain ibn Ishaq and his period. *Abh Kunde Morgenlandes, 8*:695, 1926.

Meyerhoff, M.: *The book of the ten treatises on the eye ascribed to Hunain ibn is-haq* (809-877 A.D.). Cairo, Government Press, 1928.

Meyerhoff, M.: Ibn-al-Nafis and his theory of the lesser circulation. *ISIS, 23*:100, 1935.

Miller, H.W.: The aetiology of disease in "Plato's Timaeus." *Trans Amer Philol Ass, 93*:175, 1962;

Misch, G.; *History of Autobiography in Antiquity.* 2 vols. Cambridge, Harvard University Press, 1951.

Mondeville, Henri de: *Die Anatomie des Heinrich von Mondeville.* Nach einer Handschrift der Königlichen Bibliothek zu Berlin vom Jahre 1304, ... Berlin, G. Reimer, 1889.

Mundino: *Anathomia.* 1500. (Copy in the British Library, London.)

Mondino, da Luzzi: *Anathomia, emendata per Melerstat.* Leipzig, Landsberg, 1493. (Copy in the Library of the Royal College of Physicians, London.)

Moir, D.M.: *Outlines of the Ancient History of Medicine.* Edinburgh, William Blackwood, 1831.

Moore, K.L.: Highlights of human embryology in the Koran and Hadith. *Proc 7th Saudi Medical Meeting*, 51-58, 1982.

Morsink, J.: *Aristotle on the Generation of Animals: A Philosophical Study.* Lanham, University Press of America, 1982.

Münster, L.: Allessandro Achillini, anatomico e filosofo, professore dello studio di Bologna 1463-1512. *Rivista di storia delle scienze mediche e naturali*, 24:7-22; 54-77, 1933.

Nannini, M.C.: Scienza e filosofia nell opera di Mondino de Liuzzi. *Pag Storia Med, 11*:49, 1967.

Nasr, S.H.: *Science and Civilization in Islam*. Cambridge, Harvard University Press, 1968.

Nasr, S.H.: *Islamic Science. An Illustrated Study*. World of Islam Festival Publishing Company Ltd., 1976. (Cited in Uddin, 1982.)

Obermaier, H. and Kühn, H.: *Buschmannkunst: Felsmalereien aus Südwestafrika*. Leipzig, G. Schmidt & C. Günther, 1930.

O'Brien, D.: The effect of a simile: Empedocles' theories of seeing and breathing. *J Hellen Stud, 90*:140, 1970.

Ogle, W.: *Aristotle on the Parts of Animals*. Translated with introduction and notes. London, Kegan Paul, Trench & Co., 1882.

O'Malley, C.D.: *Andreas Vesalius of Brussels 1514-1564*. Berkeley and Los Angeles, University of California Press, 1964.

O'Malley, C.D. and Saunders, J.B. de C.M.: *Leonardo da Vinci on the Human Body. The Anatomical, Physiological and Embryological Drawings of Leonardo da Vinci*. New York, Crown Publishers, Inc., 1982.

O'Neill, Y.V.: The Fünfbilderserie reconsidered. *Bull Hist Med, 43*:236, 1969.

O'Neill, Y.V.: The Fünfbilderserie — a bridge to the unknown. *Bull Hist Med*, 538, 1977.

O'Neill, Y.V.: Tracing Islamic influences in an illustrated anatomical manual. In El-Gindy, A.R. and Hassan, H.M.Z. (Eds.): *Proc 2nd Int Conf Islamic Med*, Kuwait, 1982, pp. 154-162.

Orth, E.: *Cicero und die Medizin*. Kaisersesch, Martintal, 1925.

Osler, W.: *The Evolution of Modern Medicine*. New Haven, Yale University Press, 1921.

Packard, F.R.: History of the School of Salernum. In Harrington, J. (Ed.): *The School of Salernum*. New York, Paul B. Hoeber, 1920, pp. 7-52.

Parker, R.G.: Academy of Fine Arts. Print from a copper engraving, 1578, by Cornelis Cort after a drawing by Jan Van Der Straet, 1573. *J Hist Med, 38*:76, 1983.

Parsons, E.A.: *The Alexandrian Library: Glory of the Hellenic World*. Elsevier, Amsterdam, 1952.

Peterson, D.W.: Observations on the chronology of the Galenic corpus. *Bull Hist Med, 51*:484, 1977.

Pettinato, G.: *The Archives of Ebla*. New York, Doubleday, 1981.

Peylick, J.: *Philosophiae Naturalis Compendium*. Leipzig, Melchior Lotter, 1499. (Copy in the British Library, London.)

Pilcher, L.S.: The Mondino myth: a study of the conditions attending the revival in the 13th and 14th centuries of the practical study of anatomy. In *Odium Medicum*. J.B. Lippincott Co., 1911, pp.183-208.

Potter, P.: Herophilus of Chalcedon: an assessment of his place in the history of anatomy. *Bull Hist Med, 50*:45, 1976.

Prendergast, J.S.: The background of Galen's life and activities and its influence on his achievements. *Proc Roy Soc Med, 23*:1131, 1930.

Preuss, A.: Science and philosophy in Aristotle's "Generation of Animals." *J Hist Biol 3*:1, 1970.

Qatagya, S.: Ibnul-Nafees had dissected the human body. In El-Gindy, A.R. and Hassan, H.M.Z. (Eds.): *Proc 2nd Int Conf Islamic Med*, Kuwait, 1982, pp. 306-312.

Ranke, H.: Medicine and Surgery in ancient Egypt. *Bull Hist Med, 1*:237, 1933.

Rath, G.: Pre-Vesalian anatomy in the light of modern research. *Bull Hist Med, 35*:142, 1961.

Richardson, B.W.: Vesalius, and the birth of Anatomy. *Asclepiad, 2*:132, 1885.

Richer, P.: Du role de l'anatomie dans l'histoire de l'art. *France Med, 50*:318, 1903.

Riesman, D.: *The Story of Medicine in the Middle Ages.* New York, Paul B. Hoeber, 1936.

Rollins, C.P.: Oporinus and the publication of the Fabrica. *Yale J Biol Med, 16*:129, 1943.

Ross, W.D.: *Aristotle Works.* 12 Vols. Oxford, Clarendon, 1952.

Roth, M.: *Andreas Vesalius Bruxellensis.* Berlin, G. Reimer, 1892.

Roth, M.: Vesaliana. *Virchow Arch, 141*:462, 1895.

Russel, G.A.: The anatomy of the eye: Ibn Al-Haytham and the Galenic tradition. In El-Gindy, A.R. and Hassan, H.M.Z. (Eds.): *Proc 2nd Int Conf Islamic Med*, Kuwait, 1982, p.176.

Said, H.M.: *Medicine in China.* Karachi, Hamdard Academy, 1965.

Sandelowsky, B.H.: Archaeology in Namibia. *Amer Sci, 71*:606, 1983.

Sarton, G.: *Galen of Pergamon.* Lawrence, University of Kansas Press, 1954.

Scarborough, J.: Celsus on human vivisection at Ptolemaic Alexandria. *Clio med, 11*:25, 1976.

Schipperges, H.: Mondino de Luzzi: Anatomia (1316). *Z Aerztl Fortbild (Jena), 48*:N.F.2.340, 1959.

Shanks, N.J. and Al-Kalai, D.: Arabian medicine in the Middle Ages. *J Roy Soc Med, 77*:60, 1984.

Sherrington, C.: *The Endeavour of Jean Fernel.* Cambridge University Press, 1946.

Siegel, R.E.: Theories of vision and color perception of Empedocles and Democritus; some similarities to the modern approach. *Bull Hist Med, 33*:145, 1959.

Siegel, R.E.: *Galen's System of Physiology and Medicine. An Analysis of his Doctrines and Observations on Bloodflow, Respiration, Humors and Internal Diseases.* Basel, Karger, 1968.

Sigerist, H.E.: The conflict between the 16th century physicians and antiquity. *Int Congr Hist Med, 3*:250, 1922.

Sigerist, H.E.: Alkmaion von Kroton und die Anfänge der europäischen Physiologie. *Schweiz Med Wschr, 82*:964, 1952.

Simili, A.: Un referto medico-legale inedito e autografo di Bartolomeo da Varignana. *Policlinico (Prat), 58*:150, 1951.

Simon, M.: *Sieben Bücher Anatomie des Galen.* 2 Vols. Leipzig, 1906.

Singer, C.: *The Evolution of Anatomy. A Short History of Anatomical and Physiological*

Discovery to Harvey. London, Kegan Paul, Trench, Trubner & Co., Ltd., 1925.

Singer, C.: To Vesalius on the fourth centenary of his De humani corporis fabrica. *J Anat, 77*:261, 1943.

Singer, C.: Galen's elementary course on bones. *Proc Roy Soc Med, 45*:767, 1952.

Singer, C.: *Galen on Anatomical Procedures: Translation of the Surviving Books with Introduction and Notes.* Oxford University Press, 1956.

Singer, C.: *A Short History of Anatomy and Physiology from the Greeks to Harvey.* New York, Dover Publications, Inc., 1957.

Singhal, G.D. and Guru, L.V.: *Anatomical & Obstetric Considerations in Ancient Indian Surgery.* Banaras, Banaras Hindu University Press, 1973.

Sinh Jee, B.: *A Short History of Aryan Medical Science.* Delhi, New Asian Publishers, 1978.

Siraisi, N.G.: *Taddeo Alderotti and His Pupils.* Princeton, N.J., Princeton University Press, 1981.

Sorabji, R.: Aristotle on demarcating the five senses. *Philos Rev, 80*:55, 1970.

Souques, A.: Que doivent à Herophile et à Erasistrate l'anatomie et la physiologie due système nerveux. *Bull Soc Franc Hist Med, 28*:357, 1934.

Speransky, L.S., Bocharov, V.J. and Goncharov, N.I.: The personages of Jan Stephan Van Calcar's frontispiece to Andreas Vesalius' book "On The Structure of the Human Body." *Anat Anz, 153*:465, 1983.

Steinschneider, M.: Constantinus Africanus und seine arabischen Quellen. *Arch Path Anat Phys,* 37:351, 1866.

Stierlin, H.: *The World of India.* New York, Mayflower Books, 1978.

Stroppiana, L.:' L'anatomia nel "corpus Hippocraticum." *Riv Stor Med, 7*:9, 1963.

Sudhoff, K.: *"Studien zur Geschichte der Medizin, I."* Leipzig, J.A. Barth, 1907.

Sudhoff, K.: *Ein Beitrag zur Geschichte der Anatomie im Mittelalter speziell der anatomischen Graphik nach Handschriften des 9. bis 15. Jahrhunderts.* Leipzig, J.A. Barth, 1908.

Sudhoff, K.: Ein unbekannter Druck von Johann Peyligks aus Zeite "Compendiosa capitis physici declaratio" auch "Anatomia totius corporis humani" genannt. *Sudhoff Arch, 9*:309, 1916.

Sudhoff, K.: Konstantin der Afrikaner und die Medizinschule von Salerno. *Sudhoff Arch, 23*:293, 1930.

Temkin, O.: Celsus on medicine and the ancient medical sects. *Bull Hist Med, 3*:249, 1935.

Temkin, O.: *Galenism: Rise and Decline of a Medical Philosophy.* Ithaca, Cornell University Press, 1973.

Töply, R. von: Aus der Renaissancezeit (neue Streiflichler über die Florentiner Akademie und die anatomischen Zeichnungen des Vesal). *Janus, 8*:130, 1903.

Uddin, J.: Ibn Sina's viewpoint of human anatomy. In El-Gindy, A.R. and Hassan, H.M.Z. (Eds.): *Proc 2nd Int Conf Islamic Med,* Kuwait, 1982, pp. 163-175.

Underwood, E.A.: The "Fabrica" of Andreas Vesalius. A quarter century tribute.

Brit Med J, 1:795, 1943.

Van der Ben, N.: *Empedocles: The poem of Empedocles "Peri physios". Towards a new edition of all the fragments.* Amsterdam, B.R. Grüner, 1975.

Vesalius, A.: *De Humani Corporis Fabrica Libri Septem.* Basileae, 1543.

Von Gerssdorff, H.: *Feldtbuch de Wundtartzney.* Strassburg, Johann Schott, 1517.

Von Grunebaum, G.E.: Der Einfluss des Islam auf die Entwicklung der Medizin. *Bustan,* 3:19, 1963.

Wake, W.C.: *The Corpus Hippocraticum.* Ph.D. Thesis, London University, 1952.

Walsh, J.: Galen's writings and influences inspiring them. *Ann Med Hist,* 6:1, 1934a.

Walsh, J.: Galen's writings and influences inspiring them. *Ann Med Hist,* 6:143, 1934b.

Wegner, R.N.: *Das Anatomenbildnis. Seine Entwicklung im Zusammenhang mit der anatomischen Abbildung.* Basel, B. Schwalbe & Co., 1939.

Wellman, M.: *A Cornelius Celsus: eine Quellenuntersuchung.* Berlin, Weidmann, 1913.

Wellman, M.: A. Cornelius Celsus. *Sudhoff Arch,* 16:209, 1924.

Wendt, W.E.: "Art mobilier" from the Apollo II cave, South West Africa: Africa's oldest dated works of art. *S Afr Archaeol Bull,* 31:5, 1976.

Wharton, E.: Vesalius in Zante (1564). *N Amer Rev,* 175:625, 1902.

Wilford, F.A.: Embryological analogies in Empedocles' cosmogony. *Phronesis,* 13:108, 1968.

Wilson, .G.: Erasistratus, Galen, and the pneuma. *Bull Hist Med,* 33:293, 1959.

Wischhusen, H. and Schumacher, G.-H.: Von den ersten anatomischen Lehrsektionen bis zu systematischen Präparierübungen in Rostock. *WZ Rostock,* 17:45, 1968.

Withington, E.T.: Galen's anatomy. *3rd Int Congr Hist Med,* London, 1922.

Wright, J.: A medical essay on the "Timaeus." *Ann Med Hist,* 7:117, 1925.

INDEX

A

Achillini, Alessandro, 115, 119-121, Fig. 62
Agnodice, 46
Alaca, 72
Alcmaeon, Fig. 18
  anatomical discoveries, 29
  physiological studies, 29-30
Alderotti, Taddeo
  Bologna, human dissections in, 80
Alexander, 28, 44, 45, 48, 50, 55, 59, 73, 134
Alexander the Great, embalmed body of, 44
Alexandria
  anatomical studies, 45
  decline of, 48
  dissection, 45, 134
  human skeleton in, 59
  library, 44
  vivisection, 45
Anathomia, of Mondino de Luzzi, 89-90, 93
Anatomical illustrations, 77, 90, 94, 97, 117, 126, 127, 128, 129, 130, 131, 134, 137, 157-159, 162, 163, 164, 165, 172, 173, 177, Fig. 41
Arabian medicine
  anatomical illustrations, 76-77
  anatomy, 75-77
  dissection of human body, 76
Arabian physicians, 73
Arabian scholars, 72
Arabic medicine, theory of vision, 76
Aristotle, 37, 39-43, 74, 86, 87, Fig. 20
  anatomical drawing, 39
  anatomical writings, 39
  brain, 41-42
  comparative anatomy, 42
  diaphragm, 41
  digestive tract, 40-41
  dissection of animals, 38, 42
  embryological observations, 42
  heart, 39-40
  respiratory organs and breathing, 40-41
  spleen and liver, 41
  urinary system, 41
Arnold of Villa Nova, medical writings of, 87
Art, cave and rock, 3-4
Artists, and human dissection, 101
Asclepiades of Bithynus, 50
Astrology, and anatomy, 78-79, *see* Figs. 35 & 36
Atreya, 25
Aurelius, Marcus, 60
Australopithecus afarensis, 3
Autopsy, 81, 91, Fig. 42
Avenzoar, 73
Averroes, 73, 122
Avicenna, 73, 74, 75, 86, 87, 95, 103, 158
Azzolino, 81

B

Baghdad, "House of Science", 73
Balamio, and human skeleton, 158
Bauman, Jacob, *Anatomia Deudsch*, 173, 175
Benedetti, Alessandro, anatomical books, 102
Benivieni, Antonius, and pathological anatomy, 114-115, Fig. 54

195

Berengario, Giacomo, 115, 116-118, 129, 158, Fig. 55
Bertuccio, Nicolo, successor to Mondino at Bologna, 97
Bloodletting, 120, 121, *see* Figs. 59 & 60
Bologna, 118, 120, 170
  Medical faculty, 80, 86, 91, 92, 96, 97
  Mondino, 89, 91, 92
  University of,
    dissection, 79-81, 89
    Vesalius at, 159
Boniface XIII, human dissection, 96
Boyle, Robert, 31
Brunschwig, Hieronymus
  Buch der Cirurgia, anatomical drawings in, 134
Byzantine empire, 70, 71

C

Caius, John, 126
Calcar, Jan, 165, 166, 168
Canano, G., 115, 126, Fig. 65
Canopic jars, Sons of Horus, 17, Fig. 14
Cardiovascular system, 10, 11, 21, 22, 29, 31, 34, 35, 36, 39, 40, 46, 48, 61, 63, 66, 67, 75, 76, 77, 80, 89, 95, 96, 102, 103, 104, 105, 106, 109, 117, 120, 121, 130, 131, 140, 142, 151, 168, 172, 173, 178
Celsus, 45, 51, Fig. 25
  anatomical descriptions, 55
  anatomical references, 54-55
  De Re Medicina, 54
  dissection, 55-56
  vivisection, 56
Charaka, 25
China, 20-22
  anatomical concepts, 21, 22
  body functions, 20, 21
  dissection, 20, 21, 22
  embryological concepts, 21
Cicero, *De Natura Deorum*, human body, 51-52
Columbus, Realdus, 170
Constantinus Africanus, 82, 86
Copernicus, *de Revolutionibus Orbium Coelestium*, 70
Cophon, *De Anatomia Porci*, 82
Cordoba, medical library, 73
Cyrurgia Magna, of de Chauliac, 103

D

d'Abano, Pietro, 125
dalla Torre, Marcantonio, 126
Darwin, Charles, 38, 52
da Varignana, Bartolomeo, 80, 81
da Vinci, Leonardo, 101-113, 165
  anatomical discoveries, 105-107, 112
  anatomical drawings, 102, 104, 109-113
  anatomical studies, 102-105
  demonstration of cerebral ventricles, 105
  experiments and discoveries, 105-108
  and Francesco di Melzi, 111
  human dissection, 102
  manuscripts, 111
  method of dissection, 103, 105
  musculoskeletal system, 105
  notebooks and other manuscripts, 104-105, 111-112
de Chauliac, Guy, University of Montpellier,
  teaching of anatomy at, 97, 103
De Humani Corporis Fabrica, 161-181
  Jan Calcar and other artists, 165
  outline of the work, 166-169
  plagiarism, copies and other editions, 169-175
  woodblocks, 165-166
  writing and publication of, 162-165
de Laguna, Andres, 115, 124-126, 128
de Mondeville, Henri, 80, 97
de Vigevano, Guido, 97, Figs. 43-45
de Zerbi, Gabriele, 115, 118-119
Democritus, 52
Digestive system, 7, 11, 12, 16, 17, 21, 22, 24, 25, 35, 40, 41, 46, 55, 67, 68, 75, 82, 89, 94, 95, 99, 107, 109, 117, 121, 122, 129, 130, 133, 140, 141, 142
di Hamusco, Juan Valverde, 170
Diocletian, 71
Diodorus Siculus, 16
Diogenes of Apollonia, 32
Dissection
  animals, 29, 38, 39, 42, 58, 59, 60, 64, 65, 111, 148
  human, 22, 25, 26, 33, 45, 46, 49, 50, 55, 56, 58, 60, 76, 79, 87, 89, 90, 91, 92, 93, 94, 95, 96, 97, 101, 102, 115, 117, 123, 124, 125, 126, 127, 128, 129, 149, 151, 152, 153, 155, 156, 157, 161, 170

Dryander, Johannes, 115, 128-129, 130, 131, 170, Fig. 66

E

Ebla, ancient clay tablets, 6
Egypt, 7-19
  embalming and mummification, 12-19
Egyptian, papyruses, anatomical and surgical terms in, 7-12
Empedocles
  embryological theories, 31
  physiological studies, 31
Erasistratus, 45, 47-49, Fig. 21
  anatomical and physiological studies, 47-48
  theory of disease, 48
Estienne, Charles, 127
Ethiopia, 86
Eustachius, Bartholomeus, 29, 170

F

Fabricius, 126
Fallopius, Gabriele, 126, 171, 172
Fernel, Jean, 149, 153, Fig. 89
Ferrara
  University of, Canano at, 126
Fossil discoveries, 3
Frederick II, and human dissection, 81, 82, 87
Fünfbilderserie, schematic anatomical drawings, 77, 130

G

Galen, Claudius, 28, 45, 57-69, 70, 71, 73, 74, 76, 86, 87, 89, 90, 95, 96, 102, 103, 117, 122, 123, 124, 134, 148, 149, 150, 151, 159, 160, 166, 169, 170, Fig. 26
  anatomical descriptions, 59
  anatomical knowledge, 58
  anatomical treatises, 60, 61, 62-65
  arteries and veins, 63
  brain and cranial nerves, 62
  description of bones, 62
  dissection of animals, 58, 64-65
  experimental studies, 67
  Heraclianus, 57
  Hunain ibn Ishaq and, 68
  Kühn's edition, 58
  larynx, voice production by, 68
  muscles, 62, 151
  On Anatomical Procedure, 60, 61
  On Bones, 59
  on medical scholars, influence of his writings, 60
  On the Usefulness of the Parts of the Body, 60, 61, 151
  Pelops of Smyrna, 57
  physiological system, pneumatic theory, 67
  pneumatic theory, 65
  reproductive organs, 63-64
  Satyrus, 57
  scientific achievement, 68
  spinal cord and nerve lesions, 67-68
  surgeon, 58
  treatises, 57-58
  vascular and respiratory systems, 66-67
Galenism, influence on examinations for physicians, 59
Geminus, Thomas, 173, 175, 177, *see* Fig. 102
Gerard of Cremona, translation of Rhaze's writings, 74
Gersdorf, *Feldtbuch der Wundtartzney, see* Figs. 73, 85, 86
Geynes, John
  Royal College of Physicians, Galenism, 59
Gisbertus Carbo, and Vesalius, 148, 155
Greek medicine, 27
Guenther, Johann, 115, 149, 151-153, Fig. 88
Gutenberg, 77

H

Hadiths, embryological concepts in, 72
Hadrian, 51
Haly Abbas, 73, 74, 76, 86
Harvey, William, 21, 67
Heraclianus, 57
Herodotus, 13, 16, 18
Herophilus, 45, 46, 47, 49, 55
Hippocrates, 33-37, 39, 73, 74, 87, 124, 148, 149, Fig. 19
  blood vessels, 34-35
  bones, 33, 34
  concepts of human organs, 35-36
  dissection, 33
  medical writings, 33

visceral anatomy, 34
Horus
  sons of, canopic jars in embalming, 17
Hunain ibn Ishaq, 68, 73, 74, 75
Hundt, Magnus, 116, 130, 131-132, Figs.
  82-84

I

Ibn-al-Haytham, *Kitab al-Menazir*, 76
Ibn Nafees, discovery of pulmonary circulation, 75, 76
Ibn Rabben at-Tabari, embryology, 75
Iskander, Albert Zaki, 16
India, 22-28, 86
  civilization and culture, 23
  dissection, 25-26
  Laws of Manu, 24
  relationship to Greek medicine, con-
    cept of
  human body, 27-28
  religious practices and healing, 23
  surgery, 24
  vedas, healing methods, 23
Islam, Arabian medicine, 71-77

J

Johanson, Donald, 3

K

Ketham, Johannes, 90, 130
Kitab-i-Susrud, 27
Ktesias, 27

L

Louvain
  University of, Vesalius at, 148, 153
Lucretius, 51, 52

M

Magnus, Albertus, 103
Malformations, 31
Massa, Niccolo, 115, 122-123, 151, Fig.
  64
Mediaeval anatomists, 114-146
  Alessandro Achillini, 115, 119-121, Fig.
    62
  Alessandro Benedetti, 115
  Antonio Benivieni, 115, Fig. 54
  Giacomo Berengario, 115, 116-118, 129,

Fig. 55
  Giovanni Battista Canano, 115, 126, Fig.
    65
  Johannes Dryander, 115, 128-129, 130,
    131, Fig. 66
  Charles Estienne, 115
  Johann Guenther, 115, 151-153, Fig. 88
  Magnus Hundt, 116, 131-132
  Andres de Laguna, 115, 124-126
  Niccolo Massa, 115, 122-123, Fig. 64
  Johann Peyligk, 116, 130, 131
  Jacobus Sylvius, 116, 149-151, Fig. 87
  Gabriele de Zerbi, 115, 118-119
Mesopotamia, 7
Middle ages, 70-88
Mondino, 89-99, 102, 114, 118, 119, 121,
  123, 128, 130, Figs. 37 & 38
  Anathomia, 114, 118
  anatomical findings, 95-96
  anatomical specimens, 96
  human dissection, 93-95
  Professor of Anatomy, 92, 96
Montanus, Joannes Baptista
  Galenia omnia opera, and Vesalius, 159
Montpellier
  University of, human dissection at, 87,
    97
Mudghda, 72
Mummification
  in Egypt, 12-19
  in Neanderthal period, 19
Muscular system, 22, 26, 27, 34, 35, 41,
  46, 48, 61, 62, 72, 76, 79, 89, 97, 99,
  102, 104, 105, 109, 117, 118, 125, 126,
  130, 162, 168, 170

N

Nervous system, 10, 16, 22, 29, 30, 31,
  35, 41, 42, 46, 48, 55, 61, 62, 63, 64,
  67, 68, 76, 81, 96, 100, 102, 104, 105,
  109, 112, 121, 130, 136, 141, 145, 149,
  153, 159, 162, 168, 169, 174, 179, 181
Nutfa, 72

O

Observationes Anatomicae, of Gabriele Fallopius, 171
Oporinus, Johannes, 158, 162, 164, 165,
  167, 172
Oribasius, 45, 73

P

Padua, 118, 126, 156, 162, 170, 176
Papyruses
  Edwin Smith, 7, 9-10
  Georg Ebers, 8-9
Paracelsus, 164
Paris
  University of, dissection at, 96, 149, 151, 153
Paul of Aegina, 73
Pelops of Smyrna, 57
Peyligk, Johann, 116, 130, 131, Figs. 78-81
Pharmacy, schools of, 73
Phryesen, Laurentius
  Spiegel der Artzney, anatomical illustrations in, 129, 132, 134
Plato, 39, 124
  philosophical concept of human body, 36-37
Pliny, 28, 52-54, 124
Pneuma, 32, 65-66
Pneumatic theory, Erasistratus, 48
Polybus, anatomical studies, 36, 37
Prehistoric period, 3-8
Pulmonary circulation
  Ibn Nafees, discovery by, 75
  Michael Servetus, discovery by, 75
Puteus, Franciscus, 170
Pythagoras, 29

Q

Qu'ran, embryological concepts in, 72

R

Rawlins, Thomas
  Royal College of Physicians, Galenism, 59
Raynalde, William, 137
Reisch, Gregor, *Margarita philosophica*, 122, 136, Fig. 74
Renaissance
  anatomical progress, 114, 161
  art, and anatomy, 101
  artists
    da Vinci, 101-113
    Dürer, 101
    Michelangelo, 101
    Pollaiuolo, 101
    Raphael, 101

  Signorelli, 101
  Titian, 101
  Verocchio, 101
  Bologna and Padua during, 114
Reproductive system, 10, 11, 21, 22, 27, 29, 31, 35, 36, 37, 39, 42, 55, 63, 64, 67, 72, 83, 86, 94, 96, 108, 110, 119, 125, 130, 137, 144, 171, 180
Resch, Rutger, 155
Respiratory system, 7, 11, 17, 22, 31, 32, 35, 39, 40, 48, 55, 62, 65, 66, 67, 68, 75, 77, 89, 99, 122, 142
Rhazes, 73, 148, 155, 156, 176
Riviere, Etienne, 127
Rock art
  prehistoric drawings of human hands, 4-5
  prehistoric Namibian, human figures in, *see* Figs. 3-5
Roman empire, 51-56, 57, 58, 70, 71, 77, 170
  decline of anatomy, 50
  human dissections, 58
  medical training, 50
Roslin, Eucharius
  Rosegarten, illustrations in, 137
Rostock
  University of, anatomical dissection, 92, Fig. 39
Royal College of Physicians
  Galenism, influence of, 59

S

Salerno, University of, 82-84
Satyrus, 57
Schott, Johannes, illustration of skeleton, 134, Fig. 86
Servetus, Michael, 75
Severus, 51
Skeletal system, 3, 16, 21, 22, 25, 27, 33, 34, 35, 46, 55, 59, 60, 62, 72, 76, 78, 96, 97, 105, 109, 112, 113, 117, 121, 130, 131, 143, 146, 148, 149, 153, 159, 168
Skylark, 27
Soranus, foetus *in utero*, Figs. 75-77
Stone age, 3-4
Stopius, Nicolaus, 162
Susruta, 25
  classification of bones and joints, 27
  dissection, 25-26

Greek medicine, 27
Sylvius, Jacobus, 59, 116, 123, 149-151, 170, Fig. 87

T

Tagault, Jean, 149
Talmud, 27
Tertullian, 45
Titian, 165
Tortulla, *De Mulierum Passionibus*, 86

U

Universities, early European,
  human anatomy in, 77-84, 87, 89, 96, 97, 98, *see* Fig. 39
Urinary system, 12, 16, 21, 22, 24, 35, 39, 41, 55, 82, 95, 107, 108, 122, 133, 142

V

Vagbhata, 25
Van Calcar, 157
Vedas, 23
Venus of Willendorf, 5
Vesalius, Andreas, 59, 128, 135, 161-181
  anatomical studies, 154-155, 156-157, 160
  Anatomicarum Gabrielis Falopii Observationum Examen, 172, Fig. 101
  China Root, 150, 176

court physician, 177
  De Humani Corporis Fabrica, 70, 161-181
  early life and education, 147-148
  Epitome, 162, 164, 165, 166, 167, 172
  and Fallopius, 171-172
  and Fernel, 149
  and Galenism, 159-160
  and Guenther, 151-153, 159
  Icones Anatomicae, 166
  medical education, 148
  Paraphrase of the Ninth Book of Rhazes, 148, 155, Fig. 90
  pilgrimage to Jerusalem, 177
  and Sylvius, 149, 150, 151
  Tabulae Anatomicae, 157-159
  as a teacher, 157
  at University of Padua, 156
  Venesection Letter, 162
Vespasian, 51
Vision, theory of, 76
Vivisection, human, 45, 46, 56, 117
Von Carolsfeld, Hans Schnoor, 166

W

William of Saliceto, 81
Winter, Robert, 162

Z

Zante, 177
Zodiacal man, 78, Fig. 36